PLAN OF PUBLICATION

The **"Annales Bryologici"** are published yearly in April and form a volume of about 160 pages, royal 8vo, with several illustrations.

The text is printed in either English, French, German or Latin. Articles, communications etc. should be sent to the editor FR. VERDOORN, P.O. Box 8, Leiden (Holland). The collaborators are kindly requested to send typewritten copy.

Price, per volume, 6 guilders, or bound in cloth 7.50 guilders.

A specimen copy consisting of a few sheets will be sent free on application.

MANUAL
OF
BRYOLOGY

EDITED BY

Fr. VERDOORN

in collaboration with:

DR. H. BUCH, DR. G. CHALAUD, H. N. DIXON, H. H. DU BUY, M. A. DONK, DR. H. GAMS, DR. A. J. M. GARJEANNE, PROF. DR. TH. HERZOG, DR. K. HOEFER, DR. J. MOTTE, PROF. DR. L. M. J. G. NICOLAS, P. W. RICHARDS, PROF. DR. F. VON WETTSTEIN, DR. R. VAN DER WIJK and PROF. DR. W. ZIMMERMANN

XII, 485 pages of text and 129 illustrations. Royal 8vo.
Price, bound in cloth: 20 guilders.

CONTENTS: I. Morphologie und Anatomie der Musci, von R. VAN DER WIJK — II. Morphologie und Anatomie der Hepaticae, von H. BUCH — III. Experimentelle Morphologie, von H. BUCH — IV. Germination des spores et phase protonémique, par G. CHALAUD — V. Association des Bryophytes avec d'autres organismes, par G. NICOLAS — VI. Cytologie, par J. MOTTE — VII. Karyologie, von K. HOEFER — VIII. Physiology, by A. J. M. GARJEANNE — IX. Genetik, von F. VON WETTSTEIN — X. Geographie, von TH. HERZOG — XI. Quaternary distribution, by H. GAMS — XII. Bryo-Cenology (Moss-Societies), by H. GAMS — XIII. Ecology, by P. W. RICHARDS — XIV. Classification of Mosses, by H. N. DIXON — XV. Classification of Hepatics, by FR. VERDOORN — XVI. Phylogenie. von W. ZIMMERMANN — Index of Plant-Names — Index of Authors.

ANNALES BRYOLOGICI

ANNALES BRYOLOGICI

A YEAR-BOOK
DEVOTED TO THE STUDY OF
MOSSES AND HEPATICS

EDITED BY

FR. VERDOORN

SUPPLEMENTARY VOLUME III

A. W. EVANS, A REVISION OF THE GENUS ACROMASTIGUM

Springer-Science+Business Media, B.V.
1934

A REVISION

OF THE GENUS

ACROMASTIGUM

BY

ALEXANDER W. EVANS

WITH 40 ILLUSTRATIONS

Springer-Science+Business Media, B.V.
1934

ISBN 978-94-015-2258-8   ISBN 978-94-015-3499-4 (eBook)
DOI 10.107/978-94-015-3499-4

CONTENTS

CONTENTS

INTRODUCTION

The writer's genus *Acromastigum* (10) illustrates terminal branching from both lateral and ventral segments and is the only genus of the Hepaticae with which the latter type of branching has been associated. The branches arising in this exceptional manner are flagelliform in character, and each shows at the base a narrow and incomplete underleaf. This represents the part of the segment that was not involved in the formation of the branch. In the writer's discussion of branching in the leafy Hepaticae (11, p. 23), terminal branching from ventral segments is definitely distinguished as the "*Acromastigum*" type, and its distinctive features are pointed out.

In the genus *Bazzania*, or *Mastigobryum* as it is often called, the ventral flagelliform branches are intercalary in origin and arise in the axils of underleaves. Except for this important difference the genera *Acromastigum* and *Bazzania* have much in common, and the only species of *Acromastigum* which has so far been recognized was originally referred, with some doubt, to the genus *Mastigobryum*. This species, which is now known as *A. integrifolium* (Aust.) Evans, is apparently confined to Hawaii.

Several other species, however, which are still included in the genus *Bazzania* (or *Mastigobryum*), exhibit the *Acromastigum* type of branching (see REIMERS, 26, p. 142). Two of these, *B. bancana* (Sande Lac.) Trevis. of Indo-Malaya and *B. exilis* (Lindenb.) Trevis. of South Africa, agree with *A. integrifolium* in having undivided leaves and underleaves. The others have bifid leaves and trifid underleaves, and the divisions of the leaves, in most cases, are strikingly unlike. Well-known species showing these peculiarities are *B. inaequilatera* (Lehm. & Lindenb.) Trevis. of Indo-Malaya and *B. anisostoma* (Lehm. & Lindenb.) Trevis. of Australasia and Chile. The group, of which these species are representative, was recognized, as long ago as 1845, by the authors of the S y n o p s i s H e p a t i c a r u m (14, p. 218), who

divided the genus *Mastigobryum* into three sections: A, characterized by undivided leaves; B, by bifid leaves; and C, by tridenticulate
leaves. It is the section B which corresponds in a general way with
the group under consideration. This section, in the Synopsis, includes
M. inaequilaterum, *M. anisostomum*, and four additional species, one
of which has since been transferred to the genus *Lembidium*, while
another has been reduced to synonymy.

In 1885 SPRUCE expressed the opinion that three of the species
included in section B might justly be separated as a distinct genus
(34, p. 382, footnote) but made no formal proposal to this effect. In
the following year STEPHANI referred these three species to his
section *"Inaequilatera"* of the genus *Mastigobryum* (36, p. 245) and
associated with them six other species with bifid leaves and two with
undivided leaves. His section *Inaequilatera*, therefore, is somewhat
broader in scope than the section B of the Synopsis. A few years later
SCHIFFNER followed STEPHANI's example and recognized the *"Inaequilaterae"* as a section of the genus *Bazzania* (31, p. 101).

Although STEPHANI was familiar with SPRUCE's views regarding
the three species of section B, he did not share these views (35, p. 133)
He pointed out that the *Inaequilatera* agreed closely with the sharply
defined genus *Mastigobryum* in important vegetative and sexual
characters and concluded that a generic separation would be un-
warranted. When he raised the *Inaequilatera* to subgeneric rank in
1908 (39, p. 414) he maintained the same position and listed the
characters upon which he based his conclusion. These included the
method of branching (i.e., terminal branching from the ventral
halves of lateral segments), the flagelliform character of the ventral
vegetative branches, and the ventral position of the sexual branches.
At the same time he emphasized some of the distinctive features of
the *Inaequilatera*, which he designated as an ancient phylogenetic
branch from the normal *Mastigobryum* line, and showed that the
group was now distributed over an extensive area, in certain portions
of which it had developed very characteristic types.

In the writer's opinion the presence of the *Acromastigum* type of
branching would by itself be sufficient to separate the *Inaequilaterae*
from *Bazzania*. It will be shown, however, that this important feature
is accompanied by certain histological peculiarities in the axial
organs which yield further differential characters of importance.

With the separation of the group the question at once arises whether the *Inaequilaterae* should be recognized as a distinct genus or included in the genus *Acromastigum*, necessitating a few changes in the original generic characters. Although something might be said in favor of either of these procedures the latter seems to be the more advisable. The type-species of *Acromastigum*, to be sure, differs from most of the *Inaequilaterae* in having the leaves transversely inserted, rather than obliquely inserted and distinctly incubous in their arrangement. The typical species of the *Inaequilaterae*, moreover, with bifid leaves diverge rather widely, not only from the type-species, but also from the *Inaequilaterae* with undivided, obliquely inserted leaves. At the same time even these striking differences intergrade to a certain extent and might very properly be interpreted as sectional, rather than generic, in value. This view is supported by the fact that the genus *Bazzania*, after the removal of the *Inaequilaterae*, would still include species with undivided, bidentate, and tridentate leaves.

STEPHANI, as already noted, included eleven species in the *Inaequilatera*, when he first described the group as a section of *Mastigobryum*. He divided the section into two subsections: *a*, characterized by undivided leaves; and *b*, by unequally bifid leaves. The first included *M. bancanum* Sande Lac. and one other species, the second the nine remaining species. In 1908, when he revised the *Inaequilatera* as a subgenus, he excluded *M. bancanum*, listing it among the species unknown to him, and no longer maintained the subsections *a* and *b*. By this time the number of species in the group had increased to seventeen, and STEPHANI has since proposed three additional species as new, making a total of twenty. Three of these, however, in the opinion of the writer are synonyms, leaving a residue of seventeen. If these are transferred to *Acromastigum* and added to *A. integrifolium*, *Bazzania bancana*, and *B. exilis*, the genus will contain twenty described species. In the present revision eight species are proposed as new, including two of MITTEN's manuscript species and one of SCHIFFNER's. These will increase the total to twenty-eight.

The type specimens of the published species, which are here assigned to *Acromastigum*, are preserved in various herbaria. Through the kindness of correspondents most of these type specimens have been available for examination, and the writer has studied also many of the specimens upon which other records for the species have been based. To all who have made possible the study of this valuable material the writer would here express his sincere thanks. In the citation of specimens under individual species the following abbreviations are used: A., herbarium of the Auckland Institute and Museum, New Zealand; B., herbarium of the University of Berlin; Bog., herbarium of the Buitenzorg Botanic Gardens; F., herbarium of the University of Florence (material from the DE NOTARIS collection); G., herbarium of the University of Geneva (material from the STEPHANI collection); H., Cryptogamic Herbarium of Harvard University (material from the SCHIFFNER, STEPHANI and TAYLOR collections); Herz., herbarium of TH. HERZOG, Jena; Hodg., herbarium of Mrs. E. A. HODGSON, Napier, New Zealand; L., herbarium of the University of Leiden (material from the VAN DER SANDE LACOSTE collection); Manch., herbarium of the Manchester Museum; Mich., herbarium of the University of Michigan; N. Y., herbarium of the New York Botanical Garden (including material from the MITTEN collection); P., herbarium of the Plant Research Station, Palmerston, New Zealand; V., herbarium of FR. VERDOORN, Leiden (including collections made by SCHIFFNER in 1893 and 1894); Y., herbarium of Yale University (including the private collection of the writer).

CHARACTERS OF THE GENUS A C R O M A S T I G U M, AS EMENDED

A comparison of the description given below with the original generic description of *Acromastigum* (10, p. 103) will show that the most important changes are connected with the histological structure of the axial organs, the form and method of attachment of the leaves and underleaves, and certain structural details of the cell-walls. A more detailed discussion of some of the generic features will follow the formal description.

Acromastigum Evans

Mastigobryum B. (in great part) G. L. & N. Syn. Hep. 218. 1845.
Mastigobryum [section] III. *Inaequilatera* Steph. Hedwigia 25: 245. 1886.
Bazzania section III. *Inaequilaterae* Schiffn. in Engler & Prantl, Nat. Pflanzenfam. 1³: 101. 1893.
Acromastigum Evans, Bull. Torrey Club 27: 103. 1900.
Mastigobryum [subgenus] *Inaequilatera* Steph. Bull. Herb. Boissier II. 8: 404. 1908.

Plants small to medium size: stems prostrate to suberect, loosely branched; ordinary leafy branches terminal, of the *Frullania* type (i.e., originating in the ventral halves of lateral segments), giving rise to apparent dichotomies with an undivided leaf in each fork; flagelliform branches also terminal, but of the *Acromastigum* type (i.e., originating in ventral segments), showing an incomplete and undivided underleaf at the base on one side; sexual branches also ventral, but intercalary, arising in the axils of underleaves (or, more rarely, on flagelliform branches); vegetative axial organs differentiated into a unistratose cortex composed of relatively large cells and a medulla composed of smaller cells: rhizoids simple or branched at the tips, springing from the cells of reduced leaves on the flagelliform branches and usually confined to this position on vegetative shoots, but developing also from the basal cells of fertilized female branches: leaves transversely or obliquely attached, in the latter case distinctly incubous, undivided, bidentate, or bifid, with the ventral division usually longer than the dorsal; leaf-cells with more or less thickened walls, in many cases showing a more or less evident vitta: under-leaves transversely attached, undivided or trifid: leaves of the flagelliform branches smaller than the underleaves, transversely attached, undivided or bifid: inflorescence dioicous: male branch short, simple, more or less curved; bracts smaller and more delicate than ordinary leaves, monandrous, imbricated, concave and usually bidentate at the apex; bracteoles smaller than the bracts and less concave, usually bidentate: female branch without innovations; perichaetial leaves in three or four series; increasing in size toward the archegonia, narrowly to broadly ovate, variously laciniate, lobate or dentate but usually with two main divisions; perianth in most cases unistratose, terete below, trigonous above with broad rounded keels and narrow grooves, the third keel ventral but not distinguish-able from the others, the laciniate or laciniate-ciliate mouth more or less plicate: stalk of sporophyte (so far as known) differentiated into an outer layer composed of sixteen longitudinal rows of large cells and a central core composed of smaller and more numerous cells; capsule oval, the valves four cells thick in the middle; outermost layer with nodular and often confluent thickenings, most numerous on the longitudinal radial walls; innermost layer with crowded, semiannular thickenings on the inner tangential walls; spores minutely verrucu-lose; elaters bispiral.

TYPE SPECIES: **Mastigobryum** ? **integrifolium** Aust.

The differentiation of the shoot system in *Acromastigum*, in spite of the difference in origin of the ventral vegetative branches, is essentially like that of *Bazzania*. In other words four types of branches are present: the ordinary dorsiventral vegetative branches, the radial flagelliform branches, the male branches, and the female branches. The dorsiventral branches lie approximately in a hori-

zontal or inclined plane and show a marked distinction between the
lateral leaves and the underleaves. The flagelliform branches grow
vertically downward and show no such distinction in their three
rows of small scale-like leaves. The male branches agree with the
ordinary branches in showing a distinction between bracts and
bracteoles, and the female branches agree with the flagelliform in
being essentially radial and in producing three rows of similar
perichaetial leaves. Although the distinctions between the two types
of vegetative branches in *Acromastigum* are very definite in the
majority of cases, poorly developed plants may show a tendency
for one type to intergrade into the other. It should be noted also
that the dorsiventral vegetative branches do not invariably arise
according to the *Frullania* type of branching. In very rare instances
such a branch is intercalary in origin and springs from the axil of an
underleaf. The same phenomenon is found also in *Bazzania* and is of
frequent occurrence in the paleotropic *B. vittata* (Gottsche) Trevis.
(see EVANS 12, p. 112).

The axes of the vegetative branches exhibit histological features of
unusual interest. They show, in the first place, a clear differentiation into cortex and medulla and, in the second place, a definite
arrangement of the cortical cells. In his discussion of the stem-
structure in the leafy Hepaticae, HERZOG distinguishes two types in
which a differentiation into cortex and medulla is apparent (15, p. 67)
In the first type the cortex is built up of several layers of narrow cells
with thick walls, constituting a more or less sclerenchymatous sheath
around the more delicate medulla, which is composed of larger cells
with thinner walls. In the second type the cells of the cortex are
larger than those of the medulla and cover it with a thin-walled, more
or less spongy layer, to which HERZOG ascribes the function of water-
storage. The two examples of this type which he figures are *Cephalozia connivens* (Dicks.) Lindb. (*f. 59, C*) and *Lembidium dendroides*
Carr. & Pears. (*f. 59, D*). In the *Cephalozia* the cortex is unistratose
throughout; in the *Lembidium*, unistratose on the dorsal surface and
tristratose on the ventral. The species of *Acromastigum* agree with
Cephalozia connivens in having a unistratose cortex composed of
larger cells than those of the medulla; but these cells, instead of
being thin-walled, are thick-walled. The cortex is to be looked upon,
therefore, as a xerophytic feature and undoubtedly increases the

rigidity of the axis, whatever other functions it may have. In certain robust species the thickening is more or less uniformly distributed and is so pronounced that the cell-lumina are much reduced in size (Fig. 9, B), and even in the more delicate species the outer walls are still distinctly thickened (Fig. 26, B). The medullary cells, as a rule, have much thinner walls than the cortical cells, but in certain species the walls are distinctly thickened. This condition is espcially striking in $A.$ *integrifolium*, which shows minute lumina in the medullary cells connected by narrow pits (fig. 2, A).

The cortical cells of the vegetative branches show a definite arrangement in longitudinal rows. In the majority of the species there are six of these rows on flagelliform branches and seven on ordinary dorsiventral branches. In the latter case four of the rows are dorsilateral in position and three ventral (see, for example, Figs. 4, B; 5, B; and 6, B). The number seven, unfortunately, for the dorsiventral branches, does not indicate an absolutely constant generic character. There are, for example, two species in which the number of rows is greater than seven, at least on robust branches. One of these is the generic type, the entire-leaved $A.$*integrifolium* (Fig. 2, A), the other is a species with bifid leaves (Fig. 18, B). In certain other species, moreover, poorly developed branches may have only two ventral rows, making six rows in all. In spite of these exceptions, however, the number seven is so nearly constant that it deserves especial emphasis.

On intact plants the rows of cortical cells are especially distinct in the internodes. At the nodes, where the leaves and underleaves are inserted, the rows are obscured by the presence of additional cortical cells, intermediate in size between the normal large cells and the smaller cells of the appendicular organs. These intermediate cells, as they may be called, form transverse or oblique rows. In the case illustrated (Fig. 9, C) a transverse row of six such cells is represented, indicating the line of attachment of an underleaf. A longitudinal section (Fig. 9, D) shows that the cavities of the intermediate cells are separated from the large cortical cells and from the basal cells of the underleaf by relatively thin walls. They thus faciliate exchange between the axis and the underleaf. Although a few species show more than six intermediate cells at the base of an underleaf, the number six seems to be constant in the majority of cases. Two intermediate cells thus correspond with each underleaf-division in species with trifid underleaves. In some of the smaller and more delicate species, however, the number of intermediate cells may be reduced to five or even to four. In the case of the lateral leaves the number of intermediate cells is apparently less uniform.

In the usual arrangement of the cortical cells on the dorsiventral branches it is obvious that the four dorsi-lateral rows are derived from the two lateral rows of segments cut off from the apical cell and that the three ventral rows are derived from the ventral row of segments. The arrangement represents, in fact, the persistence of a relationship which appears very early in the ontogeny of the segments. As demonstrated by LEITGEB many years ago (19, p. 5), the lateral segments in the acrogynous Jungermanniales, prior to the initiation of leaf-formation, usually consist of two external cells and one internal cell. The two external cells give rise to a leaf and to the lateral external part of the axis on one side. In most of the more robust genera and species, as cell-division proceeds, numerous anticlinal walls in various planes are laid down in the peripheral cells, and the cells of the outermost layer at maturity are, in consequence, not only numerous but irregular in their arrangement. If, however, no longitudinal walls are laid down in the peripheral layer the segment gives rise to only two rows of cells in the outermost layer, each row being derived from one of the two original cells. This is the condition found in the species of *Acromastigum* under consideration. In extreme cases neither longitudinal nor transverse walls are laid down, except the walls cutting off the leaf-cells and the intermediate cells; and, under these circumstances, an internode at maturity shows only two external cells side by side.

The sequence of the early cell-divisions in the ventral segments of the leafy Hepaticae is less uniform than in the lateral segments. This is apparently associated with the greater diversity in the size and shape of the appendicular organs and with the total absence of such organs in certain genera. So far as known, however, the first wall in a ventral segment is periclinal and cuts off an external cell from an internal cell. The external cell gives rise to the ventral external part of the axis and, in the majority of cases, to an appendicular organ of some sort. If a species produces distinct underleaves, the external cell undergoes one or more divisions by longitudinal anticlinal walls before the underleaf-rudiment makes its appearance. In the case of *Acromastigum integrifolium* the external cell divides into three cells (see Evans, 10, p. 100, text-fig. *A*), and the same thing is obviously true of the other species. Where additional longitudinal walls are laid down, as in *A. integrifolium*, the ventral internodes at maturity show more than three rows of cortical cells; where no additional longitudinal walls are laid down, as in the majority of the other species, the ventral internodes show only three rows of cortical cells.

According to Leitgeb's account of leaf-development in the *Acrogynae* the two external cells of a young lateral segment give rise to the rudiments of two leaf-lobes (19, p. 11). These develop, in species with bifid or bilobed leaves, into the divisions or lobes of the adult leaves. Even in species with bifid leaves, however, the divisions are not usually separate throughout their entire length but are coalescent toward the base, and the region of coalescence may equal or exceed in length the free portions of the divisions. This is the condition found in the species of *Acromastigum* with bifid leaves. In species with undivided leaves the coalescence between the divisions is complete, and usually no indication of lobing is apparent. At the same time an occasional leaf may be found with two minute apical denticulations, which clearly represent the vestiges of the original rudimentary lobes. Conversely, in species with normally bifid leaves, one of the divisions may be greatly reduced in size or even wholly obsolete, so that the condition found in species with undivided leaves is approximated. It will be shown below that the divisions, even where coalescent, sometimes show a marked difference in cell-structure.

The divisions of the underleaves in species with trifid underleaves may, in a similar way, be traced back to the three external cells found in a young ventral segment. Here again a basal region of coalescence is apparent, and the coalescence becomes complete in species with undivided underleaves. The scale-like leaves of the flagelliform branches are normally bifid in species with bifid leaves and undivided in species with undivided leaves. Toward the tip of a flagelliform branch the leaves become reduced in size and often fail to show the characteristic specific features. In species with bifid scale-like leaves, for example, the reduced leaves may fail to show the bifid feature clearly. In many cases the reduced leaves are associated with the production of rhizoids, which grow out directly from the leaf-cells.

The cells of the leaves and underleaves agree with those of the axial cortex in having the bounding walls more or less strongly thickened. These thickened walls show best in cross-section (Fig. 9, *D*) but can be seen also along the margins of intact leaves and underleaves, where they appear in optical section. In the majority of the species the bounding walls are plane and extend as a continuous membrane over both surfaces. In a few species, however, the walls on one of both surfaces may project as low mounds and, under these circumstances (Fig. 33, *F*), may be uniformly thickened throughout or more strongly thickened in the middle, thus forming a projecting tubercle (Fig. 35, *I*). The cuticle varies from smooth to minutely verruculose or striolate-verruculose.

The cells, in addition to the thickenings in the bounding walls, develop thickenings also in the vertical walls (i.e., in the walls separating the cell-cavities). These thickenings in some cases are in the form of distinct trigones, separated by more or less evident pits (Fig. 38, *D*), and such trigones frequently coalesce. In other cases the vertical walls appear uniformly thickened, although the cell-cavities may be rounded at the angles (Fig. 7, *C*, *D*). Both types of thickening may be present on a single leaf, and it is sometimes apparent, in the case of certain cells or groups of cells, that trigones were at first developed and that the pits between them were subsequently obliterated by deposits of cell-well substance.

In most species of *Acromastigum* the cells in the ventral part of a leaf are distinctly larger than those in the dorsal part, at least toward the base. In many cases a vaguely defined vitta is present, the cells of which are longer than broad and arranged in longitudinal rows (Fig. 29, *C*). This vitta extends upward from the base, close to the ventral margin, but soon merges imperceptibly into the adjacent tissues. In a few species the vitta is so indistinct that it would be stretching a point to speak of a vitta at all. In nearly every case the leaf-cells in the dorsal basal portion are the smallest and from this region toward the apex of the dorsal division a gradual increase in size is apparent. From the ventral base toward the apex of the ventral division there is a gradual decrease in size. As a result the two divisions in certain species are composed of subequal cells, whereas in others the condition of equality is not quite reached (Fig. 9, *H*). This description will apply not only to species with bifid leaves but

also to those with undivided leaves, in which the divisions are coalescent throughout.

It is probable that the male and female branches in *Acromastigum*,
although actually known in comparatively few species, will prove to
be fairly uniform throughout the genus. There is no evidence that the
male branches ever proliferate at the apex or that the female branches
ever develop subfloral innovations. In the male branches the bracts
and bracteoles are delicate in texture, owing to the fact that their
cell-walls are thin or only slightly thickened (Fig. 21, *J*). The
monandrous bracts are in the form of hollow sacs, with a strongly
arched keel and connivent margins (Fig. 36, *D*), and the apex is bifid
(Fig. 36, *E*) or bidentate, even in *A. integrifolium*, the only species
with undivided leaves in which the male inflorescence is known. The
bracteole, although convex from below, is not at all saccate, and the
apex is bifid (Fig. 36, *G*) or undivided, according to the condition
found in the vegetative leaves.

The leaves of the female branches are less delicate than the male
bracts and bracteoles, although the thickenings of the cell-walls are
less pronounced than in the vegetative leaves. A broadly ovate to
lanceolate-ovate form prevails throughout the genus, so far as
known, and the apex (at least in the leaves of the innermost series) is
variously laciniate or laciniate-ciliate, although in species with bifid
leaves two main laciniae or divisions are present. The perianth
conforms to the hypogonianthous type and shows three high rounded
keels in the upper part, separated by deep grooves (Fig. 11, *O*); in
the lower part the perianth is terete, and no traces of the keels are
present (Fig. 11, *N*). The contracted mouth is variously laciniate or
laciniate-ciliate (Fig. 24, *F*). The cells of the perianth are, for the
most part, in a single layer, but scattered division-walls parallel with
the surface can be demonstrated in some cases, especially toward
the base.

The writer has seen mature sporophytes in only one species of
Acromastigum. Perhaps the most distinctive features are those found
in the stalk, which consists of an outer layer, composed of large cells
arranged in eight longitudinal rows, and an internal core, composed of
smaller and more numerous cells. In the example figured (Fig. 11, *P*)
more than sixteen internal cells show in cross-section. Before the
elongation of the stalk the external cells are opaque and appear

gorged with nutritive material, while the internal cells are hyaline. The stalk just described is very similar to the stalk of *Lepidozia setacea* (Dicks.) Mitt. Here, according to DOUIN (9, p. 365, *pl. 8, f. 28*), the outer layer is composed of eight longitudinal rows of large cells, just as in the *Acromastigum*, but the internal core shows only sixteen cells in cross-section. The valves of the capsule in the *Acromastigum* are four cells thick in the middle, and the wall-thickenings agree with the type found throughout the *Acrogynae*, according to the statements of ANDREAS (1, p. 201). In the outermost layer, for example, the thickenings are in the form of low ridges, largely restricted to the longitudinal radial walls; whereas in the innermost layer, the thickenings are in the form of half-rings, restricted to the inner tangential walls.

The genus *Bazzania* differs from *Acromastigum*, not only in the intercalary origin of the flagelliform branches, but also in the histological features of the axial organs. These show a very vague differentiation into cortex and medulla (see, for example, EVANS, 12, p. 70, *pl. 13, f. 1, 2*), although the cells of the peripheral layer are a trifle larger than the interior cells and have slightly thicker walls. The cells of this layer, moreover, are more numerous than in most of the species of *Acromastigum*, and the number of longitudinal rows is not only indefinite but always more than seven. The leaves and underleaves of *Bazzania* yield a few additional differential characters. The species with tridentate leaves, for example, find no analogues in *Acromastigum*, and the lobes of the species with bilobed leaves tend to be subequal in size. The underleaves of *Bazzania* are much more diverse than those of *Acromastigum*. In many species they are subentire or variously dentate; and, even when the marginal indentations are deeper, the resulting lobes or divisions are variable in number. The underleaves, therefore, never show the definitive trifid character, which is found in the species of *Acromastigum* with bifid leaves. The scale-like leaves of the flagelliform branches, moreover, are rarely if ever bifid in *Bazzania*, whereas this is the usual condition in *Acromastigum*. A further generic difference is perhaps to be found in the stalk of the capsule. In *Acromastigum*, according to the single example noted above, the stalk has eight rows of large peripheral cells; in *Bazzania tricrenata* (Wahlenb.) Trevis., according to MÜLLER (25, p. 269, f. 76*d*), the stalk has sixteen such rows. This difference, however, can not be accepted as constant until more species have been examined.

Two other genera, *Lepidozia* and *Herpocladium*, should be mentioned in this connection. The genus *Lepidozia*, in fact, shares a number of important characters with *Acromastigum* and is doubtless closely related. Both genera, for example, bear terminal branches of the *Frullania* type, with an incomplete leaf between the branch and the higher axis; both develop their sexual organs on short ventral inter-calary branches, arising in the axils of underleaves; and both show a large-celled, unistratose cortex on the axial organs, although the cortical cells in *Lepidozia* are relatively smaller than in *Acromasti-gum* and, in most cases, arranged in more numerous longitudinal rows. The most important distinctions between the genera are derived from the vegetative organs. In *Lepidozia* the shoot-system is more or less definitely pinnate, and the branches are not only smaller than the main axis but, in many cases, limited in growth; the leaves

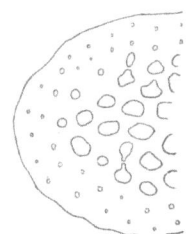

FIG. 1. *Herpocla-dium gracile* (Mont.) Steph.
Cross-section of stem, × 225. The figure was drawn from the type of *Herpocladium bidens* Mitt., col-lected by HILLE-BRAND; this species is now considered a synonym of *H. gracile*.

in most of the species have more than two lobes or divisions and the underleaves more than three; the incomplete leaf between a branch and the higher axis is, with rare exceptions, bifid; the flagelliform branches, if present at all, are nor-mally lateral; and, in the rare instances where they occupy a ventral position, their origin is intercalary. In *Acromastigum*, on the other hand, the shoot-system is not pinnate but repeatedly dichotomous (at least in appearance), and each lateral branch is essentially like the higher axis; the leaves are undivided or with only two di-visions, and the underleaves undivided or trifid; the incomplete leaf at the base of a branch is undivided; and the flagelliform branches are both ventral and terminal in origin. It may be recalled that LINDENBERG and GOTTSCHE, many years ago, looked upon *Bazzania divaricata* (Nees) Trevis., a characteristic species of *Acromastigum*, as a connecting link between the genera *Baz-zania* and *Lepidozia* (20, p. 20).

In the MITTEN Herbarium several species of *Acromastigum* are referred to the genus *Herpocladium*, and the two genera agree in having ventral flagelliform branches. The type-species of *Herpocladium*, which is known only from Hawaii, was described by MITTEN under the name *H. bidens* (24, p. 405) but is now known as *H. gracile* (Mont.) Steph. Specimens of this species examined by the writer show that the flagelliform branches are intercalary and arise in the axils of the underleaves, just as in *Bazzania*. They show also that the antheridial spikes, instead of occupying short ventral branches, occur at intervals on normal leafy branches; that the leaves and underleaves are both bifid and essentially alike; and that the axes are differentiated into a cortex, composed of about two layers of small cells with strongly thickened walls, and a medulla composed of larger cells (Fig. 1). These features indicate that the affinities of *Herpocladium* are not with *Acromastigum* and the other genera of the *Cephaloziaceae* but with the *Ptilidiaceae*, to which group SCHIFFNER has already assigned it (31, p. 106).

DESCRIPTION OF SPECIES

The differential characters utilized in distinguishing the sections and species of *Acromastigum* are drawn entirely from the vegetative organs. This will become clear from the key to the sections and from the keys to the species under sections II and IV. Sections I and III require no keys, since each includes but a single species at the present time. The sexual branches and perianths are apparently more uniform than the vegetative branches. Most of the species, however, are still known in sterile condition only.

KEY TO THE SECTIONS

1. Leaves undivided or bidenticulate at the apex; underleaves undivided   2
   Leaves bifid; underleaves trifid   3
2. Leaves tranversely attached, squarrose . .   Section I. **Squarrosa** (p. 17)
   Leaves obliquely attached, not squarrose   Section II. **Exilia** (p. 20)
3. Leaves transversely attached, subcomplicate . . . .
                                    Section III. **Subcomplicata** (p. 31)
   Leaves obliquely attached, not subcomplicate . . . .
                                    Section IV. **Inaequilatera** (p. 35)

SECTION I. SQUARROSA

Cortical cells of ordinary vegetative branches in more than seven longitudinal rows; leaves undivided, transversely attached, squarrose; underleaves undivided.

1. **Acromastigum integrifolium** (Aust.) Evans

Mastigobryum ? integrifolium Aust. Bot. Gaz. 1: 32, 1875.
Bazzania ? integrifolia Evans, Trans. Connecticut Acad. 8: 225. 1892.
Acromastigum integrifolium Evans, Bull. Torrey Club 27: 103. *pl. 1.*
    1900.

The species was based on specimens collected by Baldwin on the island of West Maui and is still known only from the Hawaiian Islands. HERZOG, in fact, lists *Acromastigum* as an endemic Hawaiian genus (16, p. 360), basing his statement on the original generic description. The following specimens of *A. integrifolium* have been examined: —

H a w a i i: West Maui, 1875, D. D. BALDWIN (Manch.), type of *Mastigobryum ? integrifolium* Aust.; Konahuanui, Oahu, 1899, C. M. COOKE, Jr. (Y.). SCHIFFNER reports the species also from the mountains behind Honolulu, Oahu, on the basis of specimens collected by the WILKES Expedition (34, p. 157).

The transversely inserted, squarrose leaves give *A. integrifolium* a very characteristic habit and will serve to distinguish it at once from the other members of the genus. The more significant characters of the plant have already been indicated in the writer's description of 1900; but the more detailed account of certain features, which is given below, will serve to emphasize still more strongly the isolated position in the genus which the species occupies.

It has already been shown that the cells of the axes have very thick walls and that the cells in the interior are longer than those of the external layers (see EVANS, 10, *f. 7*, which represents a longitudinal section through the cells). A cross-section of an ordinary vegetative branch (Fig. 2, *A*) indicates even more clearly that the external layer may be interpreted as a unistratose cortex, well differentiated from the enclosed medulla by the greater diameter of its cells. The cortical cells measure about 40 μ in both tangential and radial width, and the medullary cells average about 25 μ in diameter. The thickening of the walls, even in the medullary cells, is so pronounced that the cavities are greatly reduced in size. In the cortex the external walls measure 10—16 μ in thickness, and the distance between two adjacent cell-cavities in the medulla may be as much as 20 μ. These cavities communicate by means of small pits with narrow canals, and the middle lamellae of the cells show distinctly. In the section figured the cortex is composed of fifteen cells, but other sections of the same branch show seventeen cells. The number of longitudinal rows, therefore, although definitely more than seven, is apparently subject to some irregularity. In the flagelliform branches fewer rows are present, and the section figured (Fig. 2, *B*) shows only nine rows. The medullary cells of these flagelliform branches have thinner walls than those of the ordinary branches. The middle lamellae, although visible, are not indicated in the figure.

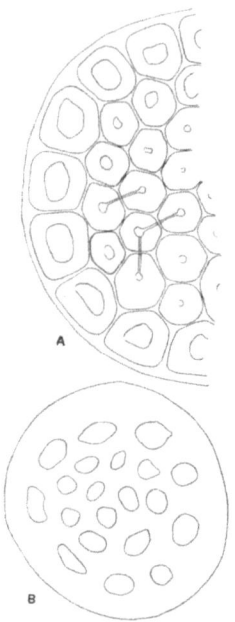

FIG. 2. *Acromastigum integrifolium* (Aust.) Evans. *A*. Cross-section of branch, × 225. *B*. Cross-section of flagelliform branch, × 225. The figures were drawn from the specimens collected by COOKE at Konahuanui, Oahu, Hawaii, in 1899.

The leaves of *A. integrifolium*, when dissected off and spread out flat, are symmetrical or nearly so and show no distinctions between the dorsal and ventral portions. They are ovate or ovate-oblong in outline and average about 0.7 mm. in length by 0.4 mm. in width. In some cases the margin is entire throughout but, as a rule, vague and irregular sinuations are present, especially toward the obtuse or subacute apex. In rare instances two low apical projections might be interpreted as the tips of the two coalescent divisions, of which the leaf is theoretically composed. In the majority of leaves, however, there is no basis for such an interpretation.

In the basal part of a leaf a broad median band is distinguishable, in which the cells are arranged in more or less definite longitudinal rows. A few of the cells in each row are longer than broad, but the majority approach an isodiametric condition. The band becomes less distinct at or above the middle, and the cells in the apical portion are more irregular in their arrangement. The same thing is true of a narrow border one to three cells wide on each side of the leaf toward the base. The median band represents a vaguely defined vitta, but the symmetry of the leaf makes it different from the vittae of species with unsymmetrical leaves. In these the vitta is ventral in position, and the band of cells separating it from the dorsal margin is, in consequence, broader than the band separating it from the ventral margin.

The leaf-cells of *A. integrifolium* average about 14 μ in width along the margin of the leaf, 18 μ in the middle, and 23 μ at the base. They are distinguished by well-developed trigones with strongly convex sides (10, *f. 5, 6*). In some of the cells the convexity is so pronounced that the trigones acquire a trifoliate outline with indentations between the convexities. Between trigones of this type and circular trigones, in which no indentations are visible, there are all possible intergradations. Coalescences between trigones are frequent but apparently never involve more than two. They may be present in longitudinal walls but are scarcely less frequent in transverse and oblique walls. The pits separating the trigones are unusually distinct and give the cell-cavities a stellate appearance.

The underleaves are somewhat smaller than the leaves and average about 0.5 mm. in length by 0.3 mm. in width. They are distinguished also by their broad, rounded or truncate apices and by the presence of apical slime-papillae or their vestiges. In some cases one or two shallow and vague indentations are present at the apex, indicating the boundaries between the three regions derived from the three external cells of the young ventral segment. The cells of the under-leaves are essentially like those of the leaves.

SECTION II. EXILIA

Cortical cells of ordinary vegetative branches in seven longitudinal rows; leaves undivided or bidenticulate at the apex, obliquely attached, neither squarrose nor subcomplicate; underleaves undivided.

1. Leaves undivided 2
   Leaves bidenticulate at the apex . . . 4. **A. bidenticulatum** (p. 26)
2. Leaves acute . . . . 2. **A. bancanum** (p. 20)
   Leaves obtuse to rounded . . . . 3. **A. exile** (p. 24)

2. **Acromastigum bancanum** (Sande Lac.) comb. nov.

Mastigobryum bancanum Sande Lac. Ann. Mus. Bot. Lugdano-Batavi 1: 301. *pl. 7, f. 1—6.* 1864.
Bazzania bancana Trevis. Mem. Ist. Lomb. 13: 414. 1877.

This interesting species was based on specimens collected on the island of Banka by VAN DIEST. These have served, not only for the original description and illustrations, but also for a series of drawings kindly communicated by Professor SCHIFFNER and for the figures in STEPHANI's unpublished Icones. The species is there included in the subgenus *Integrifolia* of the genus *Mastigobryum*, although it is not assigned to this subgenus in the S p e c i e s H e p a t i c a r u m. The following specimens have been examined:

B a n k a : Mt. Maras, 2000 ft. alt., VAN DIEST (H., L.), type of *Mastigobryum bancanum* Sande Lac.

B o r n e o : Mt. Linga, Sarawak, 1867, O. BECCARI (F.), listed by DE NOTARIS (8, p. 293); Mt. Matang, Sarawak, A. H. EVERETT (N. Y.); Mt. Bangok, Sarawak, A. H. EVERETT (N. Y.).

M a l a c c a : Mt. Ophir, Johore, H. N. RIDLEY 721 (N. Y.); same locality, 1930, F. VERDOORN 91, 131 (V., Y.).

The plants of *A. bancanum* grow in loose tufts or mats, either in pure condition or mixed with other bryophytes. They vary in color from a pale green or yellowish green to different shades of brown. The living portions are mostly 1—1.5 cm. in length, and the successive dichotomies are 2—5 mm. apart. Well-developed axes measure

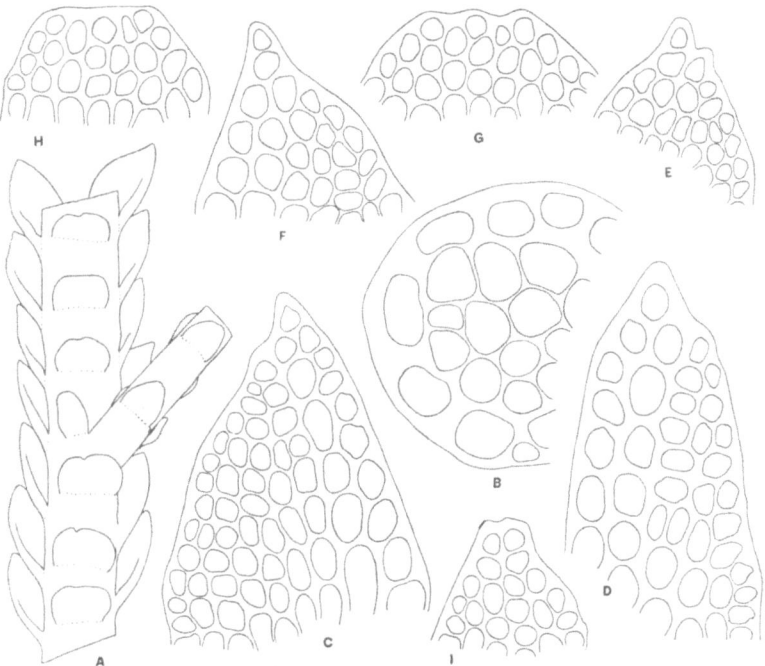

FIG. 3. *Acromastigum bancanum* (Sande Lac.) Evans *A*. Part of plant, showing dorsiventral and flagelliform branches, ventral view, × 50. *B*. Cross-section of branch, × 225. *C*, *D*. Leaves, × 225. *E*, *F*. Apices of leaves, × 225. *G*, *H*. Underleaves, × 225. *I*. Leaf of flagelliform branch, × 225. The figures were drawn from the specimens collected by RIDLEY on Mt. Ophir, Johore, Malacca.

0.18—0.2 mm. in width and 0.16—0.17 mm. in thickness; the dorsiventral compression, therefore, is relatively slight.

The cells of the clearly differentiated cortex are mostly 40—50 µ in tangential width and about 30 µ in radial width, whereas the medullary cells average about 25 µ in diameter (Fig. 3, *B*). The thickest walls are the external walls of the cortical cells, which may attain a thickness of 8—10 μ; the walls of the medullary cells are much thinner but show triangular thickenings at the angles. Owing to the close proximity of the leaves and underleaves, the seven longitudinal rows, in which the cortical cells are arranged, are difficult to demonstrate. The two dorsal rows of alternating cells are perhaps the most distinct. Between every two cells of each row, however, a leaf is attached, and the smaller intermediate cortical cells at the base of the leaf interrupt the continuity of the row. The three ventral rows are interrupted in the same way by the intermediate cells at the base of the underleaves. Since it is almost impossible to secure a cross-section in which none of the intermediate cells appear, the number of cortical cells showing is almost invariably more than seven.

The leaves (Fig. 3, A) which spread at a narrow angle, are imbricated and, in general, more or less convex. When a branch is examined from below the ventral margin toward the base appears in profile view; toward the apex, however, the convexity of the ventral part of the leaf gradually flattens out, and a narrow strip including the apex itself appears plane or nearly so. The leaves show little or no dilation at the dorsal base, where the margin meets the axis at an acute angle. They have a broadly to narrowly ovate outline (Fig. 3, C, D) and measure, when well developed, 0.25—0.3 mm. in length by 0.15—0.2 mm. in width at the base. The apex is normally acute and tipped with a single cell or with two cells in a row (Fig. 3, F). In some cases indications of a bidenticulate condtion are apparent at the apex (Fig. 3, E), but many of the leaves are entire throughout.

The cells in the ventral part of a leaf are distinctly larger than those in the dorsal part (Fig. 3, C, D), and in most leaves the large cells are arranged in three more or less definite rows, at least toward the base. Of these the submarginal row might be interpreted as the vitta, although there is little to distinguish it from the rows on each side, except at the base. Here the cells measure 30—40 μ in length by about 30 μ in width. The cells of the marginal row, of the submarginal row toward the apex, and of the third row from the margin average about 20 μ in diameter, whereas those of the dorsal part of the leaf toward the base average about 15 μ. The cell-walls, for the most part, are uniformly thickened, without evident trigones or pits, but the cell-cavities have rounded angles. In the ventral part of the leaf, where the thickening is most pronounced, the walls may be as much as 10 μ in thickness; in the dorsal part toward the margin they are rarely more than 4 μ. The cuticle is smooth.

The usual evidence of a bidenticulation is a shallow depression close to the apex on the dorsal margin of a leaf. This depression in most cases is situated between the terminal cells of the submarginal row and of the third row from the margin (Fig. 3, D) but may not be in exactly this position (Fig. 3, C). When the depression is present the terminal cell of the marginal row represents one of the teeth, the terminal cell of the third row from the margin the other. The latter is usually nothing more than a low and vaguely defined bulging (Fig. 3, C, D) but may, in rare cases, be more distinct (Fig. 3, E). The depression evidently indicates the boundary between the parts of the leaf developed from the two external cells of the young segment. The two ventral rows, therefore, would correspond with the ventral division of a leaf in a species with bifid leaves.

The underleaves of A. bancanum are distant to contiguous and closely appressed to the axis. They are considerably smaller than the leaves and measure, in most cases, only 0.07—0.08 mm. in length by 0.12—0.16 mm. in width. They are subtrapezoidal in outline, with straight or slightly bulging sides, and the broad, subtruncate apex is, in most ccses, a little narrower than the base. Many of the under-leaves are entire throughout (Fig. 3, H), but some show, at the apex, one or two slightly projecting cells, one or two slight depressions (Fig. 3, G), or other similar irregularities. The depressions, as in the leaves, indicate the boundaries between the parts of the underleaf developed from the external cells of the young ventral segment, in this case three in number. Many of the underleaves are four cells long and eight cells wide at the base, and it would probably be safe to assume that the two median rows correspond with the median division of an underleaf in a species with trifid underleaves. This would leave three rows on each side for the lateral divisions. In some cases the portion corresponding with a lateral division is more than three cells wide at the base (Fig. 3, G, at right). It may be noted in this connection that the incomplete underleaf at the base of a flagel-liform branch is, as a rule, three cells wide at the base.

The flagelliform branches are about 0.08 mm. in diameter and in many cases show clearly that the cortex is composed of six longitudinal rows of cells. The scale-like leaves, which these branches bear, are distant to subimbricated and measure, when well developed, about 0.06 × 0.5 mm. They are subovate in outline and, in most cases, narrowly truncate at the apex (Fig. 3, *I*). In some instances, however, particularly on slender branches, the apex shows two distinct denticulations.

The sexual branches of *A. bancanum* are still unknown.

Although *A. integrifolium* agrees with *A. bancanum* in its method of branching and also in its undivided leaves and underleaves, the two species differ from each other in several important characters and are apparently not closely related. In *A. integrifolium*, for example, the cortical cells of the ordinary vegetative axes are in more than seven longitudinal rows, the walls of the medullary cells are strongly thickened, the leaves are transversely attached and symmetrical in form and in cell-differentiation, and the leaf-cells show distinct trigones. In *A. bancanum*, on the contrary, the cortical cells are in seven longitudinal rows, the walls of the medullary cells are relatively thin, the leaves are obliquely attached and unsymmetrical in form and in cell-differentiation, and the leaf-cells do not show trigones. The Hawaiian plant, moreover, is considerably larger than *A. bancanum*.

3. **Acromastigum exile** (Lindenb.) comb. nov.

Mastigobryum exile Lindenb. in G. L. & N. Syn. Hepat. 217. 1845.
Bazzania exilis Trevis. Mem. Ist. Lomb. 13: 414. 1877.

The type-material of *Mastigobryum exile* was collected in South Africa by ECKLON in 1823 or a little later, and this material has served, not only for the illustrations published by LINDENBERG and GOTTSCHE (20, *pl. 2, f. 1—5*), but also for the unpublished figures in STEPHANI's Icones. Although the species has been known for so long a time, its position in the genus has been uncertain. The authors of the S y n o p s i s   H e p a t i c a r u m   placed it in section A, characterized by undivided leaves; and STEPHANI, in 1886, followed their example by including it in his section *Integrifolia* (36, p. 244). In 1908, however, he transferred it to the section *Vittata* of his subgenus *Tridentata* (39, p. 426). In the writer's opinion the true position of the species is in STEPHANI's subgenus *Inaequilatera*, i.e., in the genus *Acromastigum*, since it exhibits the *Acromastigum* type of branching. The undivided leaves would then bring it close to *A. bancanum*, and it should be noted that VAN DER SANDE LACOSTE long ago recognized the relationship between the two species (28, p. 301). A portion of ECKLON's material has been examined and may be recorded as follows:

S o u t h   A f r i c a : Cape of Good Hope, without date, C. F. ECKLON (B., N. Y.), type of *Mastigobryum exile* Lindenb. LINDEN-BERG and GOTTSCHE reported the species from the same general locality on the basis of specimens collected by DRÈGE, a contemporary of ECKLON (20, p. 10), but apparently no subsequent records have been published.

The two specimens examined by the writer, one in the GOTTSCHE Herbarium and the other in the MITTEN Herbarium, are so fragmentary that no dissections could be made. The specimen in the GOTTSCHE Herbarium consists of a single plant 1.5 cm. long, with dichotomies 1—3 mm. apart. The ordinary vegetative branches are 0.12—0.15 mm. in diameter, but the flagelliform branches are only 0.06—0.09 mm. in diameter. Each of the latter has at its base the incomplete underleaf associated with the *Acromastigum* type of branching. The ordinary branches are covered with a large-celled unistratose cortex, composed of seven longitudinal rows of cells, the external walls of which attain a thickness of 8—10 μ.

The leaves, which spread obliquely, are contiguous to subimbricated and, when well developed, measure 0.18—0.24 mm. in length by 0.12—0.15 mm. in width. They are ovate in outline and show a rounded to subcordate dilatation at the dorsal base, arching slightly beyond the middle of the axis. In most leaves the apex is broad and rounded to truncate but may be obtusely pointed. In rare instances two minute denticulations can be demonstrated in the apical region, one of which may be subacute; otherwise the leaves are entire throughout. On intact branches the leaves are more or less convex, and the ventral margin appears in profile-view when seen from below, either in the basal portion only or throughout its entire extent.

It is difficult to gain an adequate idea of the leaf-cells while the leaves are still attached. It can be seen, however, that the cell-walls are uniformly thickened without evident trigones and that a band of large cells, two or three rows wide, extends along the ventral margin, much as in *A. bancanum*. The cells of the submarginal row, or vitta, are mostly 20—30 μ in length by 20 μ in width toward the base, but their walls rarely exceed 4 μ in thickness. The marginal cells tend to be a little shorter than the submarginal, and a gradual decrease in size is apparent in passing from the submarginal row toward the dorsal margin, where the cells average only 10 μ in diameter. STEPHANI, in his Icones, states that the cuticle is verrucose.

The underleaves of *A. exile*, which lie closely appressed to the axis, are contiguous but do not overlap. They are broadly orbicular-quadrate in form, with a truncate apex, and measure about 0.1 mm. in length by 0.12—0.15 mm. in width. Occasional underleaves show one to three very short rounded projections at the apex, but the majority are either entire throughout or vaguely crenulate at the apex. Most of the underleaves are four to six cells long and eight to twelve cells wide at the base.

The leaves of the flagelliform branches are much like the under-leaves but are farther apart and measure only 0.06 × 0.08 mm., when well developed. Most of them are three or four cells in length by four to six cells in width, and their truncate apices are vaguely crenulate. Nothing is yet known about the sexual branches.

The leaves of *A. exile* are blunter and a trifle smaller than those of *A. bancanum*, the leaf-cells are smaller, and the cell-walls are thinner; otherwise the species are much alike, so far as their vegetative organs are concerned. Of course the important characters separating *A. bancanum* from *A. integrifolium* will separate *A. exile* equally well from the Hawaiian species.

4. **Acromastigum bidenticulatum** sp. nov.

Pusillum, pallide-fuscum vel flavo-fuscum, hepaticis consociatum; caules parce ramosi; folia imbricata, oblique patula, ovata, 0.25—0.3 mm. longa, 0.15—0.18 mm. lata, apica bidenticulata, margine integro; cellulae in parte ventrali circa 20 μ latae, in parte dorsali circa 10 μ latae, parietibus incrassatis; foliola remota bis subimbricata, integra, late orbiculata vel subquadrata, apice truncata; flores ignoti.

B o r n e o: Mt. Linga, Sarawak, 1867, O. BECCARI 56 in part (F.), type. The type-material of the present species, which is in the DE

NOTARIS Herbarium, is mixed with a part of the original material of *Mastigobryum echinatiforme* De Not. and with three or four other species of leafy hepatics. The packet containing the mixture is labeled, *"Mastigobryum* colla *Jungermannia* No. 20", and forms a part of No. 56. No other specimens are known.

The plants vary in color from pale brown to golden brown, and the living portions are about 1 cm. in length, with the dichotomies 1—3

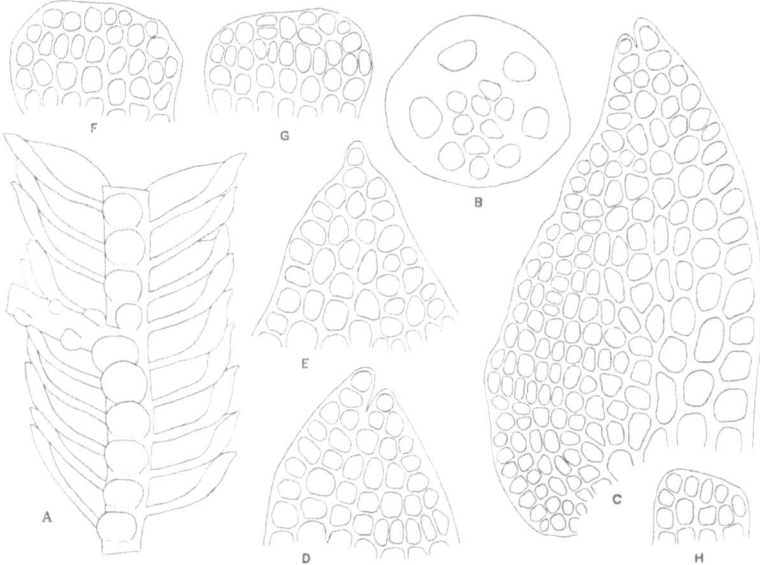

FIG. 4. *Acromastigum bidenticulatum* Evans. *A.* Part of plant, showing dorsiventral and flagelliform branches, ventral view, × 50. *B.* Cross-section of stem, × 225. *C.* Leaf, × 225. *D, E.* Apices of leaves, × 225. *F, G.* Under-leaves, × 225. *H.* Leaf of flagelliform branch, × 225. The figures were drawn from the type-material.

mm. apart. The ordinary vegetative branches are about 0.12 mm. wide and 0.11 mm. thick, and cross-sections not infrequently show seven cortical cells (Fig. 4, *B*), although the leaves are as close together as in *A. bancanum.* The cells of the four dorsi-lateral rows are 40—50 μ in tangential width and about 40 μ in radial width, but those of the three ventral rows are smaller and average only 20 μ in width; the medullary cells have an average diameter of about 16 μ. The outer walls of the cortical cells are 14—16 μ thick in the dorsi-

lateral rows and 8—10 µ thick in the ventral rows. The walls
separating the medullary cells are thin but appear thicker in places
where the thickenings at the angles coalesce. All the walls in cross-
section show a pigmentation verying from yellow to orange.

The leaves are closely imbricated and spread at an angle varying
from 45 to 75 degrees (Fig. 4, *A*). The basal portion for about two-
thirds the length of the leaf is evenly convex, and the ventral margin
appears in profile view, when seen from below. In the outer third of
the leaf the convexity is still marked in the dorsal part. A narrow
strip along the ventral side, however, becomes plane or nearly so, and
a distinct fold can be demonstrated between the convex portion and
the plane portion. When an attached leaf is examined from below the
plane portion shows particularly well, and it is evident that the
transition between the convex basal portion of the ventral side and
the plane portion is gradual. Unfortunately this transition is difficult
to illustrate. In Fig. 4, *A*, the two parallel lines at the base of the leaf
represent the ventral margin in profile view. The space between
these lines, as they diverge and then come together, represents the
plane portion. The upper boundary of this space (except at the
extreme tip) represents the fold between the convex and plane
portions, whereas the lower boundary represents the outer part of the
ventral margin in surface-view. If the transition were shown the upper
parallel line at the base of the leaf would curve outward and join the
lower boundary. It has already been shown that the leaves of *A.*
bancanum show a narrow plane strip along the ventral side in the
apical region, but this strip is more pronounced and definite in *A.*
bidenticulatum and is separated from the convex dorsal part of the
leaf by a distinct fold.

The leaves of *A. bidenticulatum* are rounded at the base and arch
to the middle of the axis or a little beyond. They are unsymmetrically ovate (Fig. 4, *C*) and measure, when well developed, 0.25—0.3
mm. in length by 0.15—0.18 mm. in width. The dorsal margin,
beyond the rounded base, extends in most cases as a straight or
slightly convex line to the apex but may show a slight concavity in
the outer part; the ventral margin is straight or nearly so to the
middle or a little beyond and then curves or bends gently forward. In
the majority of the leaves the narrow apex shows two distinct teeth,
separated by a short and narrow sinus (Fig. 4, *C, D*). The teeth are

one or two cells long and one or two cells wide at the base, and the ventral tooth projects slightly beyond the dorsal. It is only in rare instances that the dorsal tooth is obsolete, making the apex appear simply acute (Fig. 4, *E*). When an intact plant is examined from below the ventral tooth conceals the dorsal more or less completely. The bidenticulate nature of the leaves, therefore, is rarely apparent until the leaves have been dissected off and spread out flat. Aside from the apical teeth the margin is entire throughout.

Toward the base of the leaf (Fig. 4, *C*) the cells in the ventral part are considerably larger than those in the dorsal part; toward the apex, however, the disparity in size becomes less and less and finally disappears. Here, as in *A. bancanum*, the submarginal row might be interpreted as the vitta, but the cells of this row differ only slightly from those of the two adjoining rows, even in the basal region. The cells of the vitta measure 25—30 μ in length by about 20 μ in width, the cells of the marginal row are about as wide but more nearly isodiametric, those in the dorsal part toward the base average about 10 μ in diameter, and those in the apical region about 15 μ. The cell-walls throughout the leaf are thickened and the thickening usually appears uniform. Toward the dorsal margin, however, and in the apical region indications of trigones can sometimes be distinguished, although the pits between them are largely obliterated. The thickest cell-walls, which are in the vitta and vicinity, frequently attain a thickness of 6 μ. The cuticle is verruculose, except toward the base of the leaf, but the verruculae are difficult to demonstrate. They show most clearly along the dorsal margin and in the apical region.

The underleaves, which are distant to subimbricate, are distinctly convex when seen from below. They are broadly orbicular to subquadrate (Fig. 4, *F, G*) and measure, when well developed, 0.06—0.08 mm. in length by 0.07—0.1 mm. in width. On account of the convexity of the underleaves the basal cells often cover over and conceal the intermediate cells, which scarcely differ from them in size. Most of the underleaves are four or five cells long and seven to nine cells wide at the base. Although the cells tend to be arranged in longitudinal rows, the rows are often obscured by the longitudinal division of some of the cells. The underleaves have straight or slightly bulging sides and a broad truncate apex. The latter, in some cases, shows shallow and vague indentations, but usually nothing of this sort is apparent. If the underleaves are interpreted in the way suggested under *A. bancanum*, two approximately median rows of cells would correspond with the median division of a species with trifid under-leaves; two to four rows on each side would then correspond with the lateral divisions.

The scale-like leaves of the flagelliform branches are much like the underleaves but, in most cases, are only three cells long and four cells wide at the base (Fig. 4, *H*). The cells of both underleaves and scale-leaves are essentially like the leaf-cells, but exhibit little diversity in size.

The apices of the leaves in *A. bidenticulatum* bridge over, to a certain extent, the gap between species with undivided leaves and those with definitely bifid leaves. In *A. bancanum* and *A. exile*, which belong to the first category, the apices of the leaves usually show no indications of teeth; and, in the rare instances where such teeth are present, they obviously represent a deviation from the normal condition. They are comparable in this respect with the blunt lobes which are occasionally present on the leaves of other leafy hepatics with normally undivided leaves, such as *Jamesoniella autumnalis* (DC.) Steph. and *Odontoschisma prostratum* (Sw.) Trevis. In *A. bidenticula-tum*, on the other hand, the apices of the leaves are normally bi-denticulate, and the rare examples of apices without teeth represent the anomalous condition. The difference between the bidenticulate leaves of this species and the bifid leaves of the species belonging to the following sections is really a difference of degree. If the sinus between the teeth were to become deeper the bifid condition would be realized.

Although the differences in the apices of the leaves will at once separate *A. bidenticulatum* from *A. bancanum*, the two species are very similar. Both, for example, have undivided underleaves, and both show the same type of differentiation in their leaf-cells. In *A. bidenticulatum*, however, the axes are less robust than in *A. bancanum* and show a smaller number of medullary cells, the leaves spread more widely, the dorsal base is distinctly rounded, the plane apical area in the ventral part of the leaf is broader and more sharply defined, and the leaf-cells are appreciably smaller.

SECTION III. SUBCOMPLICATA

Cortical cells of ordinary vegetative branches in seven longitudinal rows; leaves bifid, transversely attached, subcomplicate; underleaves trifid.

5. **Acromastigum filum** (Steph.) comb. nov.

Bazzania filum Steph. Hedwigia 32:206. 1893.
Mastigobryum filum Steph. Spec. Hepat. 3: 538. 1909.

This interesting species from New Caledonia is known only from the type-material, a portion of which has been examined by the writer. This may be recorded as follows:

N e w   C a l e d o n i a :   without definite locality or date, G. DUPUY (H.), type of *Bazzania filum* from the STEPHANI collection.

The plants, which are unusually slender, are 3—5 mm. in length, with widely spreading dichotomies 0.5—2.5 mm. apart. The color is brownish yellow, varying to a pale yellowish at the tips. Well-developed vegetative branches are only 0.1 mm. in width by 0.9 mm. in thickness, and cross-sections (Fig. 5, *C*) not infrequently show the characteristic number of cortical cells, although the leaves are imbricated. Those of the dorsi-lateral rows measure 25—34μ in tangential width by 25—30 μ in radial width, whereas the corresponding figures for the ventral cortical cells are 15—20 μ by 15 μ. The medullary cells average about 15 μ in diameter and are therefore scarcely smaller than the ventral cortical cells. The outer walls of the dorsi-lateral cells are, in most cases, 8—10 μ thick, but those of the ventral cells are rarely more than 4 μ thick. The walls of the medullary cells are somewhat thinner, except where the thickenings at the angles coalesce.

The leaves (Fig. 5, *A*, *B*) are transversely attached, imbricated, and suberect, and the plants in consequence may, at first sight, appear leafless. As shown by Fig. 5, *D*, the dorsal division is in contact with the axis along the dorsal side but is otherwise free. This division forms an angle of about 90 degrees with the ventral division; the latter is slightly separated from the axis and appears in profile view, whether the plant be examined from above or from below. The leaves thus approach, to a certain extent, a complicate condition, even if the divisions are not actually appressed to each other. The basal part of the leaf is rounded rather than keeled. The leaves, when dissected off and spread out flat (Fig. 5, *E*, *F*) are seen to be almost symmetrical, ovate-rectangular in outline, slightly narrower at the apex than at the base, and bifid to beyond the middle with an acute sinus. They are very small, measuring only 0.1—0.15 mm. in length, 0.1—0.15 mm. in width at the base, and 0.08 —0.12 mm. in width at the apex. Both divisions are triangular and acute, and the apex is tipped either with a single cell or with a row of two cells. On many of the leaves the divisions are subequal in size, but the dorsal division may be a trifle shorter than the ventral. Each division, on well-developed leaves, is four or five cells long and three or four cells wide at the base, whereas the undivided basal portion is two or three cells high and six cells wide at the base. The margins are entire or vaguely sinuous throughout.

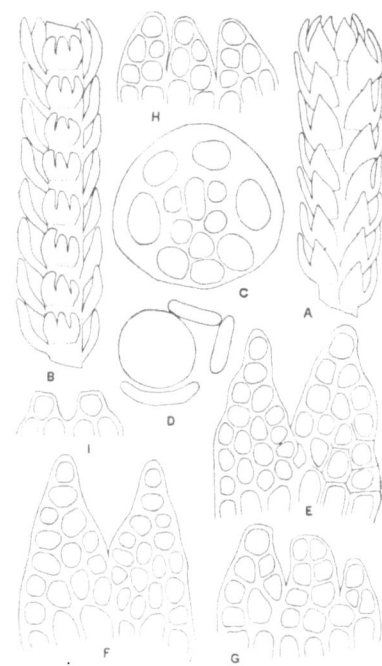

Fig. 5. *Acromastigum filum* (Steph.) Evans. *A*. Part of plant, dorsal view, × 50. *B*. Part of plant, ventral view, × 50. *C*. Cross-section of branch, × 225. *D*. Cross-section of branch, showing an underleaf and the two divisions of a side-leaf, × 100. *E*, *F*. Leaves, × 225. *G*, *H*. Underleaves, × 225. *I*. Leaf of flagelliform branch, × 225. The figures were drawn from the type-material.

Although the cells in the ventral part of the leaf are a trifle larger than those in the dorsal part, and although the ventral submarginal cells are a trifle larger than the marginal, no actual vitta is differentiated. The cells of the ventral marginal row average about 16 μ in diameter, those of the submarginal row (which may not be definite) about 22 μ in length by 18 μ in width, those of the dorsal marginal row about 14 μ in diameter, and those of the submarginal row about 18 μ in length by 14 μ in width. The cell-walls are uniformly thickened, and neither trigones nor pits are apparent. The thickest walls, which are in the ventral part of the leaf toward the base, often attain a thickness of 4—5 μ and show the middle lamellae with unusual distinctness. The cell-cavities, although sometimes more or less rounded at the angles, retain in many cases their original polygonal form. At the base of a detached leaf one or more of the intermediate cells can often be detected. These are usually six in number and are only a little larger than the basal leaf-cells. The cuticle is smooth throughout.

The underleaves of *A. filum* (Fig. 5, *B*, *G*, *H*) are approximate but apparently do not overlap, except at the tips of the branches. They are convex from below and appress themselves closely to the axis. Well-developed underleaves are subrectangular in form, with bulging sides, and measure 0.05—0.065 mm. in length by 0.075—0.085 mm. in width. The divisions, which are subparallel or slightly convergent, are separated by narrow sinuses extending half-way or a little more toward the base. Most of the divisions are two cells wide at the base, about two and one-half cells long, and tipped with a single rounded or truncate cell. The basal part of the underleaf is one and one-half cells high and, in most cases, six cells wide. Here, as in the case of the leaf, the six intermediate cells are scarcely distinguishable in size from the basal cells of the underleaf.

The flagelliform branches have a diameter of about 0.06 mm. Their bifid leaves (Fig. 5, *I*) are very minute and are composed, in most cases, of only six cells. Four of these cells form the base of the leaf, and the other two are the tip-cells of the divisions.

The specimens studied by the writer are sterile, but STEPHANI gives an account of the female branch. This arises, according to his statements, in the axil of an underleaf or from the middle of a flagelliform branch and is relatively large for the size of the plant. The perichaetial leaves are in three series. Those of the innermost, which are much larger than the others, are strongly concave and embrace the perianth in the form of a sheath. The apices of these leaves are narrow, spreading, and deeply trifid, with setaceous laciniae, and the median lacinia is the largest of the three. The leaves are composed of large, elongate-rectangular cells, with thin walls. The ovate-oblong perianth is terete below, three-keeled above, and four- or five-ciliate at the mouth, with strong and straight cilia. The cells of the perianth-wall, which are rectangular and thick-walled, measure 45 μ in length by 17 μ in width. The male inflorescence is unknown.

STEPHANI, in his original description of *A. filum*, brings out the fact that the plants were found growing on a brick-red loam, which in the tropics would indicate a xerophytic habitat. He shows further that the species is adapted to xerophytic conditions. The leaves, for example, which are reduced to a minimum, do not spread widely; the walls of their cells are strongly thickened; and the same thing is true of the walls of the superficial layer of the axes. The whole organism is thus rendered firm and rigid and might be said to have assumed the character of a desert plant.

Certain characters of *A. filum* give it a unique place in the genus. The transverse attachment of the leaves, for example, separates it from all the other known species with bifid leaves and is found nowhere else in the genus except in *A. integrifolium*. This species, however, with its undivided leaves and underleaves, is far removed from *A. filum*, with its bifid leaves and trifid underleaves. The subcomplicate relation of the leaf-divisions is another unusual feature which is not found elsewhere in the genus. Complicate and sub-complicate leaves are found in many genera of the Hepaticae, such as *Marsupella*, *Sphenolobus*, *Scapania*, *Radula*, and *Porella*. In some of these genera the complicate condition is constant and represents a generic character; in others some of the species have complicate or subcomplicate leaves, whereas others have explanate leaves; in some of the latter genera, in fact, certain species may produce either complicate or explanate leaves. The complicate condition is not necessarily an indication or relationship. It represents rather, especially when it occurs in unrelated genera, an example of parallelism, and this is the way it should be interpreted in the case of *A. filum*.

SECTION IV. INAEQUILATERA

Cortical cells of ordinary vegetative branches, with one exception, in seven longitudinal rows; leaves bifid, obliquely attached, neither squarrose nor subcomplicate; underleaves trifid, the divisions in one species reduced to crenations or even obsolete.

10. Ventral leaf-divisions, in many cases at least, longer than the dorsal divisions. 11
    Ventral leaf-divisions definitely shorter than the dorsal divisions . . 16
11. Leaf-cells with small, but usually distinct, trigones 12
    Leaf-cells with uniformly thickened walls, without evident trigones 14
12. Cortical cells of robust branches in more than seven longitudinal rows.
                                        14. **A. Colensoanum** (p. 79).
    Cortical cells even of robust branches in only seven longitudinal rows 13
13. Leaves in most cases explanate; ventral leaf-divisions, with rare exceptions, longer than the dorsal divisions . 15. **A. divaricatum** (p. 86).
    Leaves in most cases deflexed; ventral leaf-divisions in many cases shorter than the dorsal divisions . . . . . . 16. **A. laetevirens** (p. 94).
14. Leaf-divisions truncate in mature leaves . . . 17. **A. curtilobum** (p. 97).
    Leaf-divisions acute 15
15. Ventral leaf-divisions triangular, in most cases more than two cells wide at the base . . . . 18. **A. laevigatum** (p. 101).
    Ventral leaf-divisions linear, two cells wide except at the apex . . .
                                        19. **A. Cunninghamii** (p. 106).
16. Marginal cells of the dorsal divisions distinctly larger than the submarginal cells . . . . 20. **A. obliquatum** (p. 110).
    Marginal cells of the dorsal divisions subequal in size to the submarginal cells . . . . 21. **A. microstictum** (p. 115).
17. Superficial walls of leaf-cells plane or nearly so 18
    Superficial walls of leaf-cells distinctly convex, in some cases with a median tubercle 19
18. Dorsal leaf-divisions in most cases much shorter than the ventral divisions; underleaf-divisions not sharply bidentate at the apex . .
                                        22. **A. linganum** (p. 118).
    Dorsal leaf-divisions only a little shorter than the ventral divisions; underleaf-divisions in many cases sharply bidentate at the apex . .
                                        23. **A. denticulatum** (p. 125).
19. Marginal teeth of the leaves in the form of minute crenulations or denticulations, in rare instances accompanied by a few two- or three-celled teeth 20
    Marginal crenulations and denticulations accompanied by short cilia or large multicellular teeth 22
20. Vertical walls of leaf-cells uniformly thickened, without evident trigones or pits; superficial walls with or without tubercles 21
    Vertical walls of leaf-cells with large trigones and distinct pits; superficial walls with large tubercles, except toward the base of the leaf . .
                                        26. **A. echinatum** (p. 147).
21. Sinuses of the leaves in most cases less than one fourth the length of the leaves; sinuses of the underleaves about half the length of the underleaves . . . . 24. **A. inaequilaterum** (p. 129).
    Sinuses of the leaves one-fourth to one-half the length of the leaves;

6. **Acromastigum capillare** (Steph.) comb. nov.

Mastigobryum capillare Steph. Spec. Hepat. 6: 457. 1924.

This minute New Caledonian species is known only from the original material, which the writer has had the privilege of studying;
it may be recorded as follows:

New Caledonia: plateau of Dogny, without date, A. LE
RAT 420 (G.), type of *Mastigobryum capillare* Steph. STEPHANI cites
the species simply from New Caledonia; the station is taken from
the original packet.

The plants grew on bark in a depressed mat and are pale yellowish
green in color, without pigmentation of the cell-walls. The stems
rarely exceed 0.5 cm. in length, and the widely spreading dichotomies
occur at intervals of 1—3 mm. The branches of the *Frullania* type
are supplemented by intercalary dorsiventral branches arising in the
axils of the underleaves. The ordinary vegetative branches have a
diameter of about 0.1 mm. and are scarcely, if at all, compressed.
The cells of the cortical layer (Fig. 6, B) have a tangential width of
20—40 μ and a radial width of 15—30 μ. The smaller measurements
apply to the cells of the three ventral rows, and these are thus
scarcely larger in cross-section than the medullary cells, which
average about 15 μ in diameter. The bounding walls of the dorsi-
lateral cortical cells have a thickness of 10—12 μ; the medullary
cells have distinct and often coalescent thickenings at the angles,
with thin places between.

The leaves (Fig. 6, A) are imbricated and suberect, spreading in
most cases at an angle of about 45 degrees. They are distinctly
convex, and the ventral margin, at least toward the base, appears in
profile view, when a branch is examined from below. In some cases

the leaves are appressed to one another and to the axis, so that the branch appears as a leafless cylinder; in other cases the leaves are free from the axis in the outer part, as shown in the figure. The dorsal base extends to the middle of the axis but is not rounded, and the dorsal margin extends as an almost straight or slightly convex line to the apex of the dorsal division; the ventral margin is straight

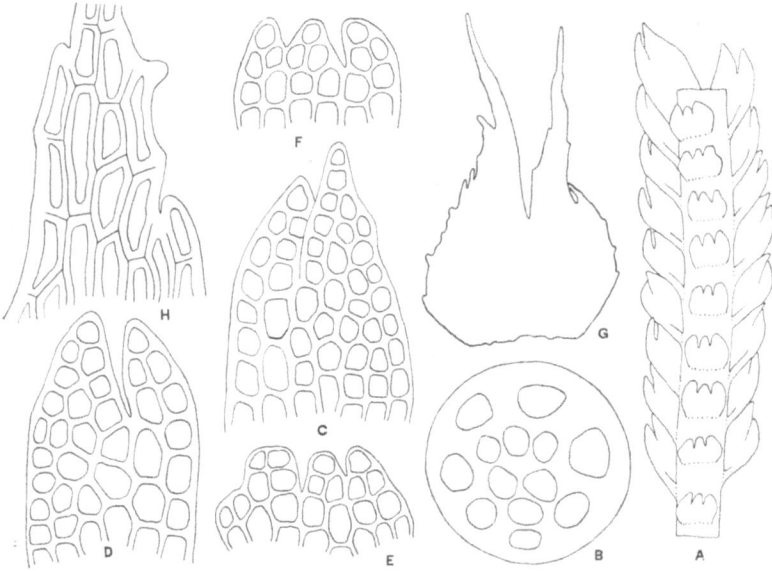

Fig. 6. *Acromastigum capillare* (Steph.) Evans. *A*. Part of plant, ventral view, × 50. *B*. Cross-section of branch, × 225. *C*, *D*. Leaves, × 225. *E*, *F*. Underleaves, × 225. *G*. Perichaetial bract from innermost series, × 50. *H*. Cells from a perichaetial bract of innermost series, × 225. The figures were drawn from the type-material.

or nearly so for more than half its length and then curves gently forward. These features are clearest in leaves which are dissected off and spread out flat (Fig. 6, *C*, *D*). Such leaves show an ovate outline, with a slight asymmetry, and measure, in well-developed examples, 0.16—0.2 in length by 0.1—0.12 mm. in width. The sinus, which is one-third to one-half the length of the leaf is acute and very narrow. The divisions, in consequence, may diverge at a small angle, lie in contact, or slightly overlap. Both divisions point obliquely forward in attached leaves, and the acute apices are tipped with a single cell or, on some of the dorsal divisions, with a row of two cells. The dorsal division is triangular, three or four cells wide at the base, and both longer and wider than the ventral division; the latter is subulate and two cells wide except at the apex. The margins of the leaves are entire throughout.

The second row of cells from the ventral margin may be interpreted as the vitta, as was done in the case of *A. bancanum* and its allies. This vitta is separated from the dorsal margin by four or five rows of cells and is composed of cells 20—30 μ in length by about 18 μ in width. The cells along the dorsal margin average about 14 μ in width and those in the dorsal division measure about 18 × 14 μ. The cell-walls are uniformly thickened, without trigones or evident pits; the middle lamellae show clearly, although not represented in the figures, and the cell-cavities are more or less angular. The thickest walls, which are in the cells of the vitta, are 4—5 μ thick, and the cuticle is minutely and faintly verruculose.

The convex underleaves (Fig. 6, *A*, *E*, *F*) are distant and appressed to the axis. They are subrectangular in general outline and measure 0.06—0.07 mm. in length by 0.1—0.11 mm. in width. The acute sinuses are one-third to one-half the length of the underleaf, and the truncate to subacute divisions are one to two cells long and one or two cells wide at the base. The truncate divisions are two cells wide at the apex, and the subacute divisions are tipped with a single cell or formed of a row of two cells. In some cases a shoulder-like bulge is present on one side of an underleaf. The basal part is two or three cells high and six to eight cells wide at the base. Two of these cells correspond, of course, with the median division and two or three with each of the lateral divisions.

The flagelliform branches have a diameter of about 0.07 mm. and bear distant scale-like leaves, which are closely appressed to the axis. These leaves are orbicular-quadrate in outline, and the largest are about 0.05 mm. in length. The apex is bidentate with unicellular, rounded to subacute teeth, or bifid with truncate to subacute divisions two cells long and one or two cells wide at the base. The basal portion is two cells high and four to six cells wide.

Although STEPHANI described the material as sterile, it includes a number of plants with unfertilized female inflorescences. Male branches, however, are apparently lacking. In the absence of peri-

anths, female branches in the genus *Acromastigum* are not altogether
favorable for study, since the perichaetial leaves fail to reach full
maturity. In the present material the branches bear three or four
series of perichaetial leaves, and those of the innermost series (Fig.
6, *G*), which are ovate in outline, measure 0.75—0.9 mm. in length
by 0.45—0.55 mm. in width. They are bifid to about the middle,
with subulate, long-acuminate divisions, each of which is tipped
with a row of two to four long cells. The sides of the leaves are irregularly dentate to ciliate. Except at the base, which appears immature, the cells (Fig. 6, *H*) are characterized by their strongly and
uniformly thickened walls, with neither trigones nor pits. They
measure 30—70 μ in length by about 14 μ in average width, the
walls may attain a thickness of 8 μ, the middle lamellae show with
unusual distinctness, and the cuticle is minutely verruculose or
striolate-verruculose. The leaves of the next outer series measure
about 0.6 $\times$ 0.35 mm., their divisions are shorter, and their margins
bear fewer teeth.

According to STEPHANI the underleaves of *A. capillare* are bifid,
but this is true only of poorly developed branches. On well-developed
branches they are almost invariably trifid. Most of the xerophytic
features which he emphasized in the case of *A. filum* are shown
equally well by *A. capillare*, although the species grows on bark
instead of on bare soil.

In its minute size and in certain characters drawn from the leaves
and underleaves *A. capillare* will bear further comparison with *A
filum*, although the two plants are here placed in distinct sections.
In both species the leaves are subequally divided and the divisions
of the underleaves are very short and truncate to subacute at the
apex. Of course the oblique attachment of the leaves in *A. capillare*,
which is characteristic of all the *Inaequilatera*, will at once distinguish
the species from *A. filum*, in which the attachment is transverse; but
this difference in attachment is associated with other differences
which are worthy of attention. It has already been shown that the
leaf-divisions in *A. filum* are at right angles to each other and that
the difference in size between the cells of the dorsal and ventral parts
of the leaf is very slight. In *A. capillare*, however, the divisions lie
one plane, or at any rate in a uniformly curved surface, and the
cells of the dorsal part of the leaf are distinctly smaller than those

of the ventral part. It may be added that the sinuses in the underleaves of *A. filum* are a little deeper than those of *A. capillare*.

7. **Acromastigum tenax** (Steph.) comb. nov.

Mastigobryum tenax Steph. Spec. Hepat. 6: 483. 1924.

In *A. tenax* another New Caledonian species is met with which is
known only from the type-material. According to the data given in
STEPHANI's Icones, where a leaf and an underleaf of the species are
figured, this material bore the number "205." The original packet,
however, which the writer has examined, bears the number "211."
The type-material may therefore be recorded as follows:

New Caledonia: summit of Mount Mou, 1908, A. LE RAT
211 (G.), type of *Mastigobryum tenax* Steph. As in the case of *A.
capillare* the station is taken from the original packet; STEPHANI
cites the species simply from New Caledonia.

The plants grew in a depressed mat on reddish soil and show a
slight admixture with some species of *Bazzania*. They vary in color
from yellowish green to brownish green, and the cell-walls are distinctly pigmented. Individual plants are 0.5—1 cm. long and branch
by false dichotomy at intervals of 1—3 mm. In addition to the
ordinary lateral branches intercalary leafy branches in the axils of
underleaves are occasionally produced. The ordinary branches are
about 0.11 mm. in width and 0.1 mm. in thickness. The cortical cells
(Fig. 7, *B*) are 20—40 μ in tangential width by 20—30 μ in radial
width, and the medullary cells average about 16 μ in diameter. The
bounding walls of the dorsi-lateral cells frequently have a thickness
of 10—12 μ, and the medullary cells show thickenings at the angles
and thin places in some of the lateral walls.

The imbricated leaves (Fig. 7, *A*) spread normally at an angle of
45 degrees or a little more. The upper surface is distinctly convex, and
the ventral margin, at least toward the base, appears in profile
view from below. In some cases, just as in *A. capillare*, the leaves
are appressed to one another and to the axis, but on most of the
plants spreading leaves are greatly in the majority. The dorsal base
is rounded to minutely subcordate (Fig. 7, *E*) and arches to the
middle of the axis or a little beyond. When the leaves are dissected

off and spread out flat they show an unsymmetrically ovate outline (Fig. 7, C, D) and measure, in well-developed examples, 0.16—0.22 mm. in length by 0.12—0.15 mm. in width. The dorsal margin extends beyond the base as an approximately straight but more or less sinuate line to the apex of the dorsal division; the ventral margin, too, may be almost straight but shows, in many cases, a more or less distinct bulge at about the middle. The acute sinus is one-

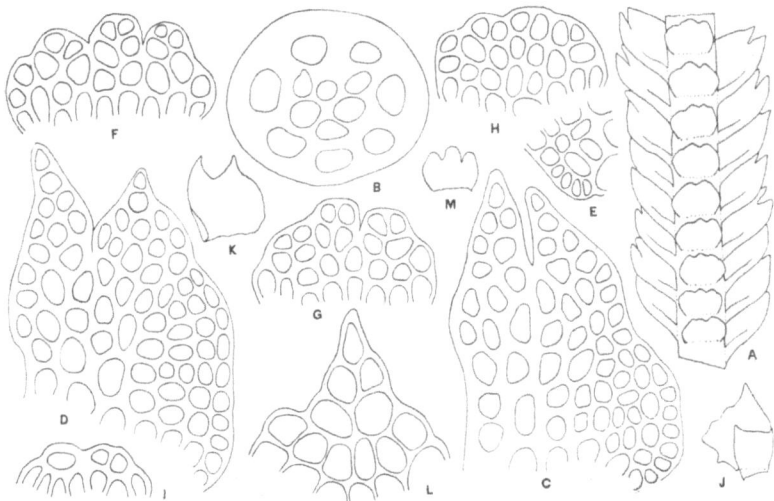

FIG. 7. *Acromastigum tenax* (Steph.) Evans. *A*. Part of plant, ventral view, × 50. *B*. Cross-section of branch, × 225. *C, D*. Leaves, × 225. *E*. Dorsal base of leaf, × 225. *F-H*. Underleaves, × 225. *I*. Leaf of flagelliform branch, × 225. *J, K*. Perigonial bracts, × 50. *L*. Apex of dorsal division of bract shown in *J*, × 225. *M*. Perigonial bracteole, × 50. The figures were drawn from the type-material.

third to one-half the length of the leaf and separates the divisions slightly. The latter are triangular and acute and are tipped with a single cell or with a row of two cells. In the best developed leaves, such as the one shown in Fig. 7, *D*, the ventral division is a little longer and a little narrower than the dorsal division, but in many of the leaves the divisions are subequal in length, or in width (Fig. 7, *C*) or in both length and width. In rare instances the ventral division is even a little shorter than the dorsal division. In a series of leaves examined the ventral divisions were two to four cells wide at the base and the dorsal divisions two to five cells wide. The margins of the leaves are entire.

The cells of the ventral part of the leaf are distinctly larger than those of the dorsal part, but the vitta must be somewhat arbitrarily defined. If the second row of cells from the ventral margin is thus interpreted, the vitta in many leaves is distinct toward the base (Fig. 7, C); in other leaves, however, this is hardly the case (Fig. 7, D). The vitta is separated from the dorsal margin by five to seven rows of cells, which show a gradual decrease in size. The cells of the vitta are 20—25 μ long and about 18 μ wide, the cells along the dorsal margin average about 10 μ in width, and those in the dorsal division measure about 16 × 12 μ. The cell-walls are uniformly thickened and may be as much as 6—8 μ thick in the vitta and vicinity. The cell-cavities have rounded angles, and the cuticle is apparently smooth throughout.

The underleaves, which are distant to contiguous, are closely appressed to the axis and thus appear convex, when a branch is examined from below. They are subrectangular in general outline, with straight or slightly bulging sides, and measure 0.06—0.07 mm. in length by 0.12—0.14 mm. in width. The most typically developed underleaves, such as the one shown in Fig. 7, F, have two narrow and acute sinuses at the apex, separating three low crenations less than two cells high, which are obtuse, rounded, or subtruncate. These crenations, which represent the divisions of species with trifid underleaves, are tipped with two cells side by side or, more rarely, with a single cell. On each side of the underleaf a rounded projection is present. Many of the underleaves, however, do not show this typical structure. Fig. 7, G, for example, represents an underleaf in which the sinus on the left is almost obsolete. In this underleaf, although the two lateral projections are indicated, the left-hand crenation scarcely projects at all, and the apex therefore appears bicrenate. In the underleaf shown in Fig. 7, H, both sinuses are almost obsolete, and the same thing is true of the crenations; the lateral projection on the right-hand side is lacking but the one on the opposite side is present. Such an underleaf might almost be described as undivided and bears a marked resemblance to the undivided underleaves of the *Exilia*. The underleaves of *A. tenax* are, in most cases, four cells high and eight to ten cells wide at the base. Two cells

thus correspond with the median crenation, and three or four cells
with each lateral crenation, together with the lateral projection.

The flagelliform branches have a diameter of about 0.06 mm. and
bear scattered, scale-like leaves, the largest of which are about 0.03
mm. in length. These leaves (Fig. 7, *I*) are broader than long, four
to eight cells wide at the base, two cells high, and bicrenate at the
apex. Here, as in the underleaves, the sinuses may be obsolete, in
which case the leaves appear undivided.

Two very short male branches have been available for study, but
the material is apparently destitute of female inflorescences. The
longer of the two male branches bore about three pairs of imbricated,
monandrous bracts. These are inflated, without being much compressed, and are bifid one-third to one-half with acute divisions.
In some of the bracts (Fig. 7, *K*) the divisions are subequal; in others
(Fig. 7, *J*) the ventral division is smaller than the dorsal. The bracts
when explanate have a broadly suborbicular form and measure
about 0.18 mm. in length by 0.24 mm. in width. The margins are
irregularly crenulate or denticulate from projecting cells, and the
cell-walls toward the tips of the divisions are slightly thickened
(Fig. 7, *L*); in the basal part of the bracts, however, the cell-walls
are thin and delicate. The bracteoles are about 0.08 mm. long and
0.13 mm. wide. One of the few examples seen was more distinctly
trifid (Fig. 7, *M*) than the ordinary underleaves, another was tricrenate, and a third was bicrenate. It is probable that the bracteoles
on a longer male branch would be more uniform.

The xerophytic features which are so pronounced in *A. filum* and
A. capillare are present also in *A. tenax*, and it is interesting to note
that the three New Caledonian species, although so distinct from one
another, are all adapted to the same type of environment. In its
marked pigmentation *A. tenax* differs from *A. capillare* and agrees
with *A. filum*, and the two species agree further in growing on a
reddish soil, an unusual habitat for species of *Acromastigum*. It will
be sufficient, however, to compare *A. tenax* with *A. capillare*, since
most of the characters separating the latter species from *A. filum*
are found also in *A. tenax*.

Aside from the difference in color *A. tenax* differs from *A. capillare*
in a number of characters drawn from the leaves and underleaves. In
A. tenax, for example, the ventral leaf-division is longer than the

dorsal in well-developed leaves; the contrast in size between the cells of the vitta, which have an average width of 18 μ, and those along the dorsal margin, which have an average width of only about 10 μ is very marked; each leaf as a rule is composed of more than sixty cells; and the underleaves are tricrenate, rather than bifid, with the sinuses and crenations often more or less obsolete. In *A. capillare*, on the other hand, the dorsal leaf division is longer than the ventral in well-developed leaves; the contrast in size between the cells of the vitta, which also have an average width of 18 μ, and the cells along the dorsal margin, which have an average width of 14 μ, is less marked than in *A. tenax*; each leaf, as a rule, is composed of fewer than fifty cells; and the underleaves are distinctly bifid even if the divisions are short.

The underleaves of *A. tenax* are of unusual interest because they serve to bridge over the gap between the trifid underleaves of the typical *Inaequilatera* and the undivided underleaves of the *Exilia*. They indicate, moreover, that there is some justification for the attempt to interpret the underleaves of the latter section in terms of coalescent divisions.

It has already been implied that trifid underleaves in *Acromastigum* probably preceded undivided underleaves in the evolution of the genus. On the basis of this assumption the change from the trifid to the undivided condition would have been brought about by an extension of the process of coalescence until the three divisions had completely lost their individuality. In *A. tenax* the coalescence has not proceeded to this extreme limit, and the three divisions are still indicated, in many cases at least, by apical crenations.

8. **Acromastigum aurescens** sp. nov.

Pusillum, laxe caespitosum vel muscis et hepaticis consociatum, nitidum, flavo-viride bis fusco-viride; caules parce ramosi; folia imbricata, subrecte patula, ovata, 0.25—0.3 mm. longa, 0.15—0.2 mm. lata, bifida, lobis triangulatis, acutis, margine integro; cellulae in parte ventrali circa 20 μ latae, in parte dorsali circa 12 μ latae, parietibus incrassatis, trigonis nullis; foliola dissita, trifida, lobis brevibus truncatis vel subacutis; flores ignoti.

M a l a c c a: Mt. Ophir, Johore, April, 1930, Fr. Verdoorn 36, 159 (V., Y.). No. 36 may be designated as the type. No other specimens are known.

The plants grow in loose depressed mats, either by themselves or in company with other bryophytes, and present a more or less glossy appearance, especially when dry. The color varies from a yellowish green, through various shades of orange, to a brownish green. Individual plants are 0.5—1 cm. in length, and the dichotomies are 1—3 mm. apart. The ordinary vegetative branches, when well developed, have a width of 0.1—0.12 mm. and a thickness of 0.09 —0.11 mm. The cortical cells (Fig. 8, *B*) are 25—30 μ in tangential width by 20 —25 μ in radial width, and the medullary cells have an average diameter of 15 μ. The cell-walls throughout are distinctly pigmented, and those bounding the cortical cells on the outside may attain a thickness of 8—10 μ. The walls of the medullary cells are much thinner but show deposits of thickening at the angles.

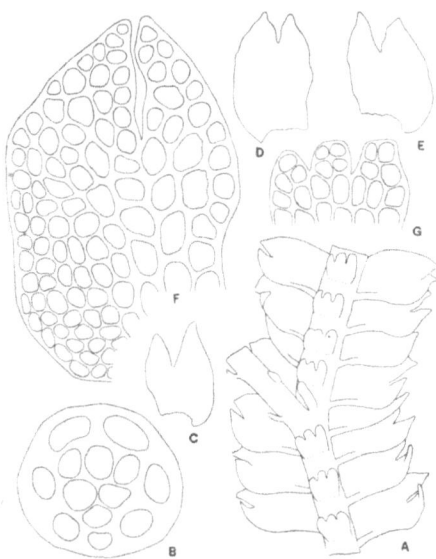

FIG. 8. *Acromastigum aurescens* Evans. *A*. Part of plant, showing dorsiventral and flagelliform branches, ventral view, × 50. *B*. Cross-section of stem, × 225. *C-E*. Leaves, × 50. *F*. Another leaf, × 225. *G*. Underleaf, × 225. The figures were drawn from the type-material.

The leaves (Fig. 8, *A*) are rather closely imbricate and spread at an angle of 70 to 90 degrees. The upper surface is more or less convex, especially in the ventral part; and the ventral margin, at least to about the middle, appears in profile view, when seen from below. On some of the leaves, in fact, the entire ventral margin as far as the apex of the ventral division, appears in profile view, and the same thing may even be true of the lower margin of the dorsal division. In some cases both divisions are more or less decurved or deflexed, but otherwise the leaves lie approximately in a single plane. Well-developed leaves (Fig. 8, *C-F*) are 0.25—0.3 mm. in length by 0.15—0.2 mm. in width and, when explanate, are ovate in general outline, with a slight assymmetry. The dorsal base is rounded and extends to about the middle of the axis. Beyond the base the margin continues as a convex line to the tip of the dorsal division. The ventral margin is, in general, straight or somewhat convex but may show a shallow concavity toward the base. The divisions (Fig. 8, *F*), which are triangular and acute, are tipped either with a single cell of with a row of two cells. The ventral division is usually narrower than the dorsal but is not necessarily shorter. It may, in fact, equal or even slightly exceed the dorsal in length. In a series of leaves examined the ventral divisions were three or four cells wide at the base, while the dorsal divisions were five to eight cells wide. The acute sinus is about one-third the length of the leaf. In some cases the sinus is so narrow that the divisions lie almost in contact with each other; in other cases the sinus is wider and the divisions may diverge up to an angle of perhaps 45 degrees. The margin is entire throughout.

The vitta is not very definite (see Fig. 8, *F*). It is usually one or two cells wide at the base but is scarcely distinguishable beyond the middle of the leaf. Between the vitta and the ventral margin are one or (more rarely) two rows of cells; between the vitta and the dorsal margin, five to eight rows. The cells of the vitta are 25—30 μ long and 20 μ wide, those along the dorsal margin average about 12 μ in width, and those in the divisions about 15 μ in diameter. The cell-walls are uniformly thickened but show, in some cases, indications of trigones, the pits between which have been obliterated by subsequent deposits of thickening. The thickest walls, which are in the vitta and vicinity, may attain a thickness of 6 μ. The cuticle is minutely verruculose or striolate-verruculose, but the markings are difficult to demonstrate except along the margin.

The underleaves (Fig. 8, *A*) are distant and appressed to the axis, showing a slight convexity from below. Well-developed examples measure 0.07—0.08 mm. in length by 0.08—0.1 mm. in width. Their outline (Fig. 8, *G*) is broadly subquadrate, and their sides usually bulge a little. The narrow and acute sinuses do not extend quite to the middle. The divisions, which are often subequal, are, in most cases, two or three cells wide at the base and two cells long. The apices are either truncate and two cells wide or subacute and tipped with a single cell. The basal portion of the underleaves is usually six to eight cells wide and two cells high. It may, however, appear to be three cells high, since the intermediate cells at the base are scarcely larger than the cells of the underleaf.

The diameter of the flagelliform branches varies from 0.06 mm. to 0.09 mm. The distant scale-leaves are very small, the largest observed being 0.03 mm. long and 0.045 mm. wide. They are bifid one-third to one-half with an acute sinus, and the divisions are one or two cells long. The apices are either truncate or tipped with a rounded cell. The basal portion in well-developed leaves is four cells wide and one or two cells high. The sexual branches are unknown.

The present species is closely allied to *A. tenax*, agreeing with it in color, in the general form of the leaves and of the leaf-divisions, and in the uniformly thickened walls of the leaf-cells. It exhibits, moreover, a number of xerophytic features, although these are less pronounced than in the New Caledonian plant. The slightly larger size of *A. aurescens* and the more widely spreading leaves will help to distinguish it from *A. tenax*, but the most important differential characters are those derived from the underleaves. In *A. aurescens* these are definitely trifid, whereas in *A. tenax* they vary from tricrenate to undivided.

9. **Acromastigum anisostomum** (Lehm. & Lindenb.) comb. nov.

Jungermannia anisostoma Lehm. & Lindenb. in Lehmann, Pug. Plant. 6: 57. 1834.

Jungermannia atrovirens Tayl. Jour. Bot. 3: 388. 1844. Not *J. atrovirens* Dumort. Syll. Jung. 51. 1831.

Mastigobryum atrovirens Tayl. in G. L. & N. Syn. Hepat. 218. 1845.

Mastigobryum anisostomum Lehm. & Lindenb. in G. L. N. Syn. Hep. 219. 1845.

Bazzania anisostoma Trevis. Mem. Ist. Lomb. 13: 414. 1877.

Bazzania Mooreana Steph. Hedwigia 33: 1. 1894.

Mastigobryum Mooreanum Steph. Spec. Hepat. 3: 539. 1909.

Mastigobryum chiloënse Steph. Kungl. Svenska Vetensk.-Akad. Handl. 46[9]: 59. *f. 22, e—h.* 1911.

The original specimens of *Jungermannia anisostoma* were collected by MENZIES on South Island, New Zealand, and those of *J. atrovirens* by HOOKER on Auckland Island, which lies about five degrees south

of New Zealand. Although the authors of the S y n o p s i s H e p a t -
i c a r u m recognized the validity of the second species, under the
generic name *Mastigobryum*, on page 218 of their work, they reduced
it to synonymy under *M. anisostomum* on page 717, which was
published two years later. STEPHANI's *Bazzania Mooreana* was based
on specimens collected by J. B. MOORE at „Spreut" River, Tasmania.
According to RODWAY (27, p. 75) this species represents „the robust
form" of *B. anisostoma*, and the writer is inclined to agree with him
in considering the two species synonymous. Another species which
is here reduced to synonymy is *Mastigobryum chiloënse* Steph.,
which was based on Chilean specimens collected by SKOTTSBERG.
Excellent figures of *A. anisostomum* have been published by LIN-
DENBERG and GOTTSCHE (20, *pl. 5, f. 1—6*), who cite both MENZIES'
and HOOKER's specimens. The figures of the species in STEPHANI's
Icones were drawn from HOOKER's Auckland Island specimens, those
of *Bazzania Mooreana* from MOORE's original material, and those of
Mastigobryum chiloënse (which supplement the published figures)
from SKOTTSBERG's original material. The following specimens of *A.
anisostomum* have been examined by the writer:

A u c k l a n d I s l a n d: without definite locality, 1840, J. D.
HOOKER (B., H., N. Y., Y.), type of *Jungermannia atrovirens* Tayl.

C h i l e: Halt Bay, Straits of Magellan, 1868, A. CUNNINGHAM 186
(N. Y.); Guiatecas Islands, 1897, P. DUSÉN (H.), cited by STEPHANI
(38, p. 50); San Pedro Island, Chiloé, 1908, C. SKOTTSBERG (G.), type
of *Mastigobryum chiloënse* Steph. MASSALONGO (21, p. 242) reports
the species from Mt. Sarmiento and the Brecknock Peninsula, Tierra
del Fuego, on the basis of specimens collected by SPEGAZZINI; and
STEPHANI cites it from Molyneux Sound, on the basis of specimens
collected by DUSÉN (38, p. 50). In a later publication he reported *A.
anisostomum* from San Pedro Island, Chiloé, and *Mastigobryum chiloënse* from Hale Cove, as well as from San Pedro Island (40, p. 59,
60), all on the basis of specimens collected by SKOTTSBERG.

N e w Z e a l a n d: Dusky Sound, South Island, without date, A.
MENZIES (B.), type of *Jungermannia anisostoma* Lehm. & Lindenb.;
Point Preservation, South Island, without date, C. LYALL (N. Y.),
cited by MITTEN (23, p. 524); without definite locality, South Island,
1874, H. KRONE (H.); Mt. Moehan, Coromandel, North Island, 1930,
L. B. MOORE 432 (Hodg., Y.); Little Barrier Island, North Island,

1933, W. M. HAMILTON 159 (Hodg., Y.); Westland, South Island, without date, R. HELMS 4935 (P., Y.); Point Pegasus, Stewart Island, without date, L. COCKAYNE (A., Y.), cited by COCKAYNE as *Bazzania Mooreana* (6, p. 144; see also HERZOG, 16, p. 372).

T a s m a n i a : North Spreut River, 1893, J. B. MOORE 663 (H.), type of *Bazzania Mooreana*; Mt. La Perouse, without date, A. OLD-FIELD (H.), cited by Stephani as *Mastigobryum Mooreanum* (39, p. 540).

The species has been listed also from Lone Cove River and Ball's Head Bay, New South Wales, Australia, by CARRINGTON and PEAR-SON (5, p. 75), on the basis of specimens collected by WHITELEGGE; and from Antipodes Island, southeast of New Zealand, by STEPHANI (39, p. 539), who does not give the collector's name.

The plants grow in depressed mats, either in pure colonies or mixed with other bryophytes, and present many xerophytic features. Except close to the apex, they are deeply pigmented with yellowish brown or reddish brown, and the axial organs may become almost black with age. The stems are longer than in any of the preceding species, with the exception of *A. integrifolium*, and frequently attain a length of 3—5 cm., with the acutely spreading dichotomies 5—20 mm. apart. Well-developed axes are 0.18—0.25 mm. in width and 0.12—0.2 mm. in thickness. The cortical cells (Figs. 9, *B*; and 11, *A*) are 40—80 μ in tangential width by 40—60 μ in radial width, and the medullary cells average about 27 μ in diameter. The walls of the cortical cells, in accordance with the xerophytic habit of the plants, are very strongly thickened, and the bounding walls may attain the unusual thickness of 25—35 μ. The walls of the medullary cells, except for the pits, are 6—8 μ thick, and triangular thickenings are

FIG. 9. *Acromastigum anisostomum* (Lehm. & Lindenb.) Evans. *A*. Part of plant, showing dorsiventral and flagelliform branches, ventral view, × 40. *B*. Cross-section of stem, × 225. *C*. Surface-cells of stem at insertion of underleaf, × 225. *D*. Longitudinal section of stem, showing base of underleaf, × 225. *E*. Cells from base of leaf, ventral side, × 225. *F*, *G*. Dorsal bases of leaves, × 225. *H*. Apex of leaf, × 225. *I*. Lateral division of under-leaf, × 225. *J*, *K*. Leaves of flagelliform branch, × 225. *L*, *M*. Perichaetial leaves of innermost series, × 40. *N*. Perichaetial leaf of second series, × 40. *O*. Cells from basal margin of perichaetial leaf, × 225. *P*, *Q*. Laciniae from mouth of perianth, × 225. The figures were drawn from the type-material of *Jungermannia atrovirens* Tayl.

present at the angles. The walls throughout are distinctly pigmented.

The leaves (Fig. 9, *A*) are distant to loosely imbricated and vary in color from a pale yellowish to a pale brownish. In many of the leaves the ventral side toward the base is more deeply pigmented than the rest and stands out as a more or less clearly defined deep yellow or orange area. The leaves spread at an angle of 60—80 degrees and in most cases are deflexed in the apical portion. In the basal portion the upper surface is more or less convex, especially toward the ventral side, and the ventral margin toward the base may appear in profile view, when a branch is examined from below. When dissected off and spread out flat the leaves show an unsymmetrically ovate outline and may be either straight or falcate (Figs. 10, *A—C*; and 11, *B—F*). Well-developed leaves are 0.6—0.9 mm. in length by 0.25—0.4 mm. in width. They are rounded to subauriculate (Figs. 9, *F*; and 11, *I*) at the dorsal base, arching a little beyond the middle of the axis, and the dorsal margin extends outward, to the apex of the dorsal division, as a straight, slightly convex, or slightly concave line. The ventral margin, which is straight or concave throughout its entire length, develops in rare cases a slight basal dilatation. The divisions, which are acute and tipped with a single cell or with a row of two or three cells, are separated by an acute sinus which is one-third to one half the length of the leaf, measured from the tip of the ventral division to the base. On most of the leaves the ventral division is distinctly longer than the dorsal and may, in extreme cases, be three or four times as long. The dorsal division (Figs. 9, *H*; 10, *G—I*; and 11, *J*) is triangular in form, but the ventral division varies from triangular to narrowly subulate (Figs. 9, *H*; 10, *J, K*; and 11, *K*). There is thus in some leaves a marked dissimilarity in form between the two, whereas in other leaves this dissimilarity is more or less eliminated. In a series of leaves examined the dorsal divisions were three to nine cells wide at the base and the ventral divisions three to five cells wide. The longest ventral divisions observed were sixteen cells in length. The margin is entire except close to the dorsal base, where one or two low and irregular projections, associated with slime-papillae, can be detected (Figs. 9, *F, G*; and 11, *I*).

In the ventral portion of the leaf the cells are distinctly larger than in the dorsal portion, and this applies not only to the cells below the sinus (compare Figs. 9, *E*; 10, *D—F*; and 11, *G*, with Figs 9, *F*; and

11, I), but also to those in the divisions (Fig. 9, H). The vitta is more distinctly marked than in most of the other species and is three or four cells wide at the base. It is separated from the ventral margin by one to four rows of cells and from the dorsal by twelve to sixteen rows. The cells in the basal part of the vitta are 25—50 μ long and 20—30 μ wide, those between the vitta and the dorsal margin average about 12 μ in diameter, those at the base of the dorsal division are 12—15 μ wide, and those at the base of the ventral division are about 20 μ wide. The cell-walls are everywhere strongly thickened. This is clearly apparent, not only in the walls separating the cell-cavities, but also in the walls along the margin, which appear in optical section and thus give a clear idea of the thickness of the walls bounding the cells above and below. These marginal walls are about 4 μ thick along the dorsal margin but may be as much as 10 μ thick along the ventral margin; the vertical walls in the vitta may likewise be as much as 10 μ thick. The cells of the vitta show more or less distinct trigones, with concave, straight, or convex sides, but the pits between them, especially in the longitudinal walls, may become partially or wholly filled up by deposits of cell-wall substance, so that extensive coalescences of trigones result. In the transverse walls the pits are more persistent. Except in the region of the vitta the walls appear uniformly thickened, with only vague indications of trigones, although the cell-cavities may acquire circular or elliptical outlines. The cuticle is smooth throughout.

The underleaves (Figs. 9, A; and 11, L), which are distant and slightly convex, lie parallel with the axis without being appressed to it. This is due to the fact that they extend at right angles to the axis at the very base and then curve abruptly forward (Fig. 9, D). Well-developed underleaves measure 0.01—0.15 mm. in length and 0.17—0.3 mm. in width. Their outline is broadly elliptical to sub-rectangular, and the line of attachment (Fig. 9, C) is considerably narrower than the base, which bears on each side a rounded to sub-auriculate expansion (Figs. 9, I; 10, L; and 11, M). The sinuses are acute and extend to about the middle, separating the divisions slightly, at any rate on detached underleaves. The divisions are broad and rounded, truncate, or slightly retuse at the apex. On some of the underleaves the divisons are subequal; on the others the median division is narrower than the lateral divisions. In the majority of cases the median division is two cells wide at the apex, four cells long, and four cells wide at the base, whereas the lateral divisions are

Fig. 10. *Acromastigum anisostomum* (Lehm. & Lindenb.) Evans. *A-C.* Leaves × 50. *D.* Cells from ventral margin of a fourth leaf, × 225. *E.* Cells from ventral margin of leaf shown in *A*, × 225. *F.* Cells from base of leaf shown in *B*, ventral side, × 225. *G.* Dorsal division of fourth leaf, × 225. *H.* Dorsal division of a fifth leaf, × 225. *I.* Dorsal division of a sixth leaf, × 225. *J.* Ventral division of fourth leaf, × 225. *K.* Ventral division of sixth leaf, × 225. *L.* Lateral and median divisions of an underleaf, × 225. The figures were drawn from the specimens collected by Cockayne on Stewart Island.

four to six cells wide at the base. On some underleaves a low, shoulder-like projection is present on one or both sides.

The flagelliform branches are 0.1—0.12 mm. in diameter, when well developed, and bear distant scale-like leaves, oblong to subquadrate in form and 0.07—0.1 mm. in length. They are normally bifid about one-third (Fig. 9, *J*), with a narrow sinus and rounded divisions, which are tipped with a single cell or with a row of two cells. Each division bears a hyaline papilla or its vestige at the apex. In poorly developed leaves, such as the one shown in Fig. 9, *K*, each papilla is situated between two cells, instead of on a single cell. In such a case the sinus is nothing more than a shallow, lunulate depression, and the divisions are truncate and exceedingly short.

The writer has not seen the male inflorescence of *A. anisostomum*, but STEPHANI has described it and has figured a male branch, together with a detached perigonial bract, in his Icones. According to his account the male branches are capitate and the crowded, monandrous bracts are in four pairs. The bracts are further described as conduplicate-concave and incised-bilobed one-third, with unequal, short-acuminate lobes. His figure of an explanate bract shows unequal lobes, but his figure of an entire male branch shows in most cases equally bilobed bracts. The bracteoles are neither described nor figured.

A female branch and perianth have been figured and briefly described by LINDENBERG and GOTTSCHE (20, p. 19, *pl. 5, f. 7*), and the writer has found a few examples of these parts in the material at his disposal. The perichaetial leaves are in three series and increase rapidly in size upward. Those of the innermost series (Fig. 9, *L, M*) are ovate and measure about 1.2 × 0.5 mm. They are subequally bifid about half their length with long-attenuate divisions and a narrow sinus. The divisions are tipped with a row of two to five elongate cells; and the margins are either entire or bear a single cilium-like tooth on one or both sides. The margins show in addition a number of scattered slime-papillae or their vestiges (Fig. 9, *O*), and these may be associated with slight irregularities of various types. The leaves of the next outer series (Fig. 9, *N*) are much the same, but those of the outermost series are only 0.3—0.35 mm. long and 0.2—0.3 mm. wide.

The perianths are fusiform and attain a length of 3 mm. and a diameter 0.9 mm. when well developed. The basal part (Fig. 11, *N*) shows the usual circular outline in cross-section and the part above the middle (Fig. 11, *O*) the usual tricarinate condition, with rounded keels and deep grooves. The contracted plicate mouth is surrounded by about twelve slender laciniae, which are three to six cells wide at the base and tipped with a cell-row composed of two to four narrow cells (Fig. 9, *P, Q*). The laciniae are 0.15—0.3 mm. in length, and their margins are sparingly and irregularly crenulate or denticulate from projecting cells. Throughout the entire extent of the perianth the cells are 20—50 μ long; in the laciniae they are 12—18 μ wide, and below the laciniae their average width is about 20 μ. The cell-walls are thickened and show distinct trigones, and the cuticle is smooth or, in the upper part of the perianth, faintly striolate.

The sporophyte of *A. anisostomum* has been described in connection with the characters of the genus. Fig. 11, *P*, represents a cross-section of the stalk, but the valves of the mature capsules were so badly shriveled that they did not show the thickenings of the cell-walls very clearly. The spores are about 12 μ in diameter and the elaters 15—18 μ in width in the thickest part.

One of the figures of *A. anisostomum* in STEPHANI's unpublished Icones shows a branch in which the leaf-divisions are exceedingly irregular and, with few exceptions, truncate or rounded at the apex. The material from which these figures were drawn shows numerous branches of this character, but the leaves owe their appearance to age and weathering, which have brought about an erosion of the apices. The figure, therefore, gives a somewhat misleading idea of the species. Two other figures, showing detached leaves are more satisfactory. The leaves in these have acute divisions and thus agree with the published descriptions of the species.

According to the older writers the ventral leaf-division in *A. anisostomum* is wider than the dorsal; according to STEPHANI (39, p. 539) the ventral division is narrowly triangular, whereas the dorsal is broadly triangular and only half as long. He does not actually state, therefore, whether the ventral division is wider or narrower than the dorsal, although the latter is perhaps implied. The plants studied by the writer show a wide range of variation in the relative width of the divisions. In a few of the leaves the ventral division is the wider of the two; in a larger number the divisions are subequal in width; but in the majority of the leaves the ventral division is slightly narrower than the dorsal. In all probability the last condition is the most typical, and deviations from this condition may be due to a tendency of the dorsal division to be stunted in its growth. A similar tendency will be noted in *A. exiguum* and may be associated with the fact that the dorsal division develops later than the ventral.

STEPHANI's figure of *Mastigobryum Mooreanum* in his Icones shows leaves in which both the divisions are acute but in which the ventral division is several times as long as the dorsal. In some of the leaves the ventral division is wider at the base than the dorsal, in others narrower. The leaves shown in STEPHANI's figure agree on the whole with the leaves shown in Fig. 11, *B, C*. The latter were drawn from OLDFIELD's Tasmanian specimens, which STEPHANI had referred to *M. Mooreanum* and which are essentially like MOORE's original specimens. OLDFIELD's specimens differ in several respects from the usual forms of *A. anisostomum*. The leaves, for example, are different in form, as a comparison of Fig. 11, *B, C*, with Fig. 11, *D—F* will show; but the differences in cell-structure are even more striking. In the usual forms of *A. anisostomum*, as shown by Figs. 9, *E*, and 11, *H*, the vitta is separated from the ventral margin by two or more rows of smaller cells; in OLDFIELD's specimens, on the other hand, as shown by Fig. 11, *G*, the vitta is separated from the ventral margin by a single row of cells, which are only a little smaller than those of the vitta.

Although these various differences might at first seem significant, they are bridged over by COCKAYNE's specimens from Stewart Island, which were referred by STEPHANI to *M. Mooreanum*. Some of the leaves in these are comparable in form with those of OLDFIELD's specimens, as a comparison of Fig. 10, *A*, with Fig. 11, *B, C*, will show, although other leaves (Fig. 10, *B, C*) show the form more usual for the species. It is in the cell-structure, however, that the gradation between the two extremes is most convincing. In Fig. 10, *D*, a group of cells along the ventral side of a leaf is shown, in which the vitta is separated from the margin by a single row of cells. This figure is thus comparable with Fig. 11, *G*, which represents the almost invariable condition in OLDFIELD's specimens. In Fig. 10, *E*, however, some of the marginal cells have undergone division; and in Fig. 10, *F*, the process has gone still farther, so that (except at two places) the vitta is separated from the margin by two rows of cells. This figure is not essentially different from Figs. 9, *E*, and 11, *H*, which were drawn from the more usual forms of *A. anisostomum*. The writer feels justified, therefore, in considering the differential characters of *Bazzania Mooreana* inconstant. It clearly represents a form of *A. anisostomum*, but whether it represents a "robust" form or not is less evident. The specimens are certainly no more robust than some of the more usual forms of the species.

STEPHANI's original description of *Mastigobryum chiloënse* makes no mention of the pigmentation of the cell-walls and does not compare the species with any of its allies. The description, however, applies very closely to *A. anisostomum*, except for the statement that the divisions of the leaves are obtuse. Since his *f. 22, e*, shows a leaf with obtuse divisions, it might seem at first that this character, if at all constant, would be sufficient to separate *M. chiloënse* specifically from *A. anisostomum*. Unfortunately the type-material, which the writer has been able to examine, does not support this view. Wherever the leaves are well developed the divisions are acute, and it is only on poorly developed and stunted leaves that the divisions may be more bluntly pointed. Apparently STEPHANI himself reached the conclusion that obtuse divisions were of exceptional occurrence. At any rate the detached leaf represented in his Icones has acute divisions, and his revised description of the species (40, p. 457) states definitely that the ventral divisions are acute. Since the type-material is deeply pigmented the writer is convinced that *Mastigobryum chiloënse* should be considered a simple synonym of *A. anisostomum*.

FIG. 11. *Acromastigum anisostomum* (Lehm. & Lindenb.) Evans. *A*. Cross-section of branch, × 225. *B-F*. Leaves, × 50. *G*. Cells from base of leaf shown in *B*, ventral side, × 225. *H*. Cells from base of a sixth leaf, ventral side, × 225. *I*. Cells from base of leaf shown in *B*, dorsal side, × 225. *J*. Dorsal division of leaf shown in *B*, × 225. *K*. Ventral division of leaf shown in *B*, × 225. *L*. Underleaf, × 50. *M*. Underleaf, × 225. *N*. Cross-section of perianth below the middle, × 40. *O*. Cross-section of another perianth above the middle, × 40. *O*. Cross-section of stalk of sporophyte, × 225. *A-C*, *G*, and *I-M* were drawn from the specimens collected by OLDFIELD on Mt. La Perouse, Tasmania; *D*, *E*, *H* and *N-P*, from specimens collected by HELMS at Westland, South Island, New Zealand, No. 4935; and *F* and *L*, from specimens collected by MOORE on Mt. Moehan, Caromandel, North Island, New Zealand, No. 432.

The distinct pigmentation of the cell-walls in this and the following species will separate them at once from all the other known *Inaequilatera* except *A. tenax* and *A. aurescens*. These two species, however, are so much smaller than *A. anisostomum*, that there is little

possibility of confusion. At the same time attention may be called to a few additional differential characters. In *A. tenax* and *A. aurescens*, for example, the walls of the leaf-cells are uniformly thickened, showing neither trigones nor pits; and the underleaves are not expanded at the base. In *A. anisostomum*, on the other hand, the walls show distinct trigones in the vitta and vicinity; pits are evident in this region, at least in the transverse walls; and the underleaves are distinctly expanded at the base.

10. **Acromastigum brachyphyllum** sp. nov.

Mediocre, gracile, flavum bis flavo-fuscum, laxe caespitosum; caules ad 6 cm. longi, parce ramosi; folia dissita bis contigua, oblique patula, orbiculato-ovata, 0.35—0.45 mm. longa, 0.25—0.4 mm. lata, bifida, lobis triangulatis, acutis vel obtusis, margine integro; cellulae hic et ibi bistratosae, in parte ventrali 20—30 μ latae, in parte dorsali circa 10 μ latae, parietibus incrassatis; foliola dissita, trifida, lobis latis, rotundatis bis truncatis; flores ignoti.

New Zealand: Paparoa Range, South Island, 3000 ft. alt., without date, R. HELMS 4948 (P., Y.). Known only from the type-collection.

The material of *A. brachyphyllum*, although scanty, indicates that the plants grew in loose tufts or mats. The color of the stems, except in the youngest parts, varies from yellow to golden brown, and the leaves are distinctly paler. The pigmentation, therefore, is much the same as in *A. anisostomum* but somewhat less pronounced. The plants are long and slender, in many cases attaining a length of as much as 6 cm., and the dichotomies, which diverge at an acute angle, are 1—2 cm. apart. Well-developed branches are 0.2—0.25 mm. in width and 0.18—0.22 mm. in thickness. The cortical cells (Fig. 12, *B*) have a tangential width of 40—60 μ and a radial width of 30—50 μ, whereas the medullary cells have an average diameter of about 25 μ. The bounding walls of the cortical cells may attain a thickness of 20—30 μ; the walls between the medullary cells, except for the more or less evident pits, are 4—6 μ thick, and show triangular thickenings at the angles.

The leaves (Fig. 12, *A*), which are distant to contiguous, spread obliquely at an angle of 45—60 degrees. The upper surface is convex,

especially along the ventral side, and the ventral margin, in consequence, appears in profile view, either throughout its entire extent
or toward the base only, when a branch is examined from below.

FIG. 12. *Acromastigum brachyphyllum* Evans. *A*. Part of plant, ventral view,
× 50. *B*. Cross-section of branch, × 225. *C*, *D*. Leaves, × 50. *E*. Cells from
base of a third leaf, ventral side, × 225. *F*. Dorsal base of a fourth leaf,
× 225. *G*. Apex of a fifth leaf, × 225. *H*, *I*. Leaf-cells in cross-section, showing
bistratose condition, × 300. *J*. Leaf-cells in surface view, showing
bistratose condition, × 300. *K*. Underleaf, × 50. *L*. Lateral division of a
second underleaf, × 225. The figures were drawn from the type-material.

No cases have been observed in which the apical portion was deflexed.
Well-developed leaves measure 0.35—0.45 mm. in length by 0.25—
0.4 mm. in width and show an unsymmetrically orbicular-ovate
outline (Fig. 12, *C*, *D*), when dissected off and spread out flat. They
are distinctly rounded at the dorsal base and may even be cordate

or subauriculate (Fig. 12, *F*), arching a little beyond the middle of the axis. Beyond the base the margin continues as a convex line for a considerable distance and then gradually straightens out or even becomes slightly concave in the outer part. The ventral margin is variable but pursues, in most cases, an approximately straight course to about the base of the ventral division, where it curves or bends gently forward. The divisions are triangular and acute to obtuse at the apex, which is tipped with a single cell or, more rarely, with a row of two cells (Fig. 12, *G*). The ventral division is longer than the dorsal and in many leaves is also broader. In some of the leaves, however, the dorsal division equals or exceeds the ventral in width. In a series of leaves examined the ventral divisions were four or five cells wide at the base and five to seven cells long, whereas the dorsal divisions were five or six cells wide and four to six cells long. The acute sinus is one-fifth to one-fourth the length of the leaf, measured from the tip of the ventral division to the base. The margin is entire throughout, except at the dorsal base, where one or two irregular crenations or denticulations, associated with slime-papillae, may be present.

The cells in the ventral part of the leaf, including the ventral division, are larger than those in the dorsal part, but the transition from the larger to the smaller cells is gradual. The same thing is true of the transition between the large cells of the vitta and the smaller cells on each side. The vitta, in fact, is not clearly defined (Fig. 12, *E*). A row of larger cells, to be sure, can be distinguished three or four rows from the ventral margin, and this may be interpreted as the vitta. The continuity of this row, however, may be interrupted here and there by the presence of a longitudinal wall, and some of the cells are no longer than broad, although others show the characteristic elongate form. The cells of the vitta, as thus interpreted, are 30—40 μ in length by 25—30 μ in width; the cells between the vitta and the dorsal margin average about 10 μ in diameter; and the cells in the ventral division about 18 μ. The cell-walls are distinctly thickened at the angles, but the trigones have, as a rule, concave sides and are often confluent. This is particularly true of the longitudinal walls in the vicinity of the vitta, so that definite pits between cell-cavities are largely restricted to the transverse walls. The thickest walls sometimes attain a thickness of as much as 10 μ; such walls can be seen in the vitta and along the margin, as well as on both surfaces of a leaf. The cuticle is smooth throughout.

A remarkable feature of the leaves in *A. brachyphyllum* is the occurrence of regions in which the cells are in two layers. This is brought about by the appearance of oblique walls in the cells (Fig. 12, *H*) or of walls parallel with the leaf-surfaces (Fig. 12, *I*). When a leaf is examined from the surface the bistratose arrangement of the cells can, in many cases, be detected by the fact that the boundaries of the cells in the one layer do not correspond with those in the other (Fig. 12, *J*). The bistratose regions are small in area, and each one involves a pair of cells or a group of three to five cells. They may appear almost anywhere in the leaf but are most frequent in the dorsal portion in the vicinity of the median line.

The underleaves (Fig. 12, *A*) are distant and are essentially like those of *A. anisostomum*. They are, in other words, slightly convex when seen from below and bear a rounded to subauriculate basal expansion on each side. This is associated with the fact that the line of attachment is only half as long as the width of the underleaf. Well-developed examples are 0.1—0.14 mm. long and 0.16—0.22 mm. wide and show a broadly elliptical outline (Fig. 12, *K*). The narrow acute sinuses extend nearly or quite to the middle; and the divisions, which converge slightly on attached underleaves, are broad and rounded to subtruncate at the apex (Fig. 12, *L*). In most cases one or both of the lateral divisions are broader than the median division, and in some cases a shoulder-like expansion is present on one or both sides. The divisions are four or five cells long and four to seven cells wide at the base.

The flagelliform branches have a diameter of about 0.1 mm., when well developed, and bear scattered scale-like leaves, the largest of which are about 0.1 mm. in length. Some of the leaves are shortly bifid, with obtuse to rounded divisions separated by a distinct sinus, but other leaves appear truncate at the apex, owing to the fact that no distinct sinus is present. Similar leaves have been described for *A. anisostomum*. The sexual branches are unknown.

There is a close relationship between *A. brachyphyllum* and *A. anisostomum*. Both species are characterized by a distinct pigmentation and by strongly thickened cell-walls, particularly in the cortical cells of the main axes, and are also in close agreement with respect to the characters derived from the underleaves. The leaves, however, yield important differential characters. These, in *A. brachyphyllum*, are shorter and relatively broader than in *A. anisostomum*, the sinus is shorter (with respect to the length of the leaf), and the divisions are more bluntly pointed, in many cases being obtuse rather than acute. The leaf-cells, too, show differences. In *A. brachyphyllum*, for example, the vitta is less clearly marked than in *A. anisostomum*, and the trigones in the vicinity of the vitta are less distinct and rarely show convex sides. Even more important as a differential character is the occurrence of bistratose regions in the leaves of *A. brachyphyllum*. In *A. anisostomum* the leaves are unistratose throughout. A very few cases have been observed, to be sure, in which the tips of the ventral divisions appeared to be bistratose, but this appearance is due to the fact that the tips in question are too strongly convex to be flattened out by pressure.

11. **Acromastigum echinatiforme** (De Not.) comb. nov.

Mastigobryum echinatiforme De Not. Mem. Accad. Sci. Torino II. 28: 302. *pl. 30.* 1874.
Bazzania echinatiformis Trevis. Mem. Ist. Lomb. 13: 414. 1877.

The original material of *A. echinatiforme* was collected by BECCARI in Borneo, and the species has since been recorded from Amboina. The writer is able to report it from two additional islands in the Indo-Malayan region. The following specimens have been examined:

A m b o i n a: without definite locality or date, J. E. TEYSMANN (B., H., N. Y., Y.), cited by STEPHANI (39, p. 537); Salhaetoe, 1889, G. KARSTEN (H.), cited by SCHIFFNER (30, p. 258). The figures in STEPHANI's Icones were drawn from TEYSMANN's specimens.

B o r n e o: Mt. Linga, Sarawak, 1867, O. BECCARI 56 (F.), type of *Mastigobryum echinatiforme* De Not.

P h i l i p p i n e s: Infanta, Province of Tayabas, Luzon, 1909, C. B. ROBINSON 9413 (N. Y.).

S u m a t r a: summit of Dĕlĕng Baroes, Karoland, 1927, H. H. BARTLETT 8489*c* (Mich., Y.).

According to HERZOG (16, p. 349) the species is one of the characteristics bryophytes of the middle Indo-Malayan region.

The plants of *A. echinatiforme* grow in depressed mats and are mixed, in many cases, with other bryophytes. In the Sumatra material the accompanying species are *Bazzania loricata* (R. Bl. & N.) Trevis. and *B. erosa* (R. Bl. & N.) Trevis. The color of the *Acromas-*

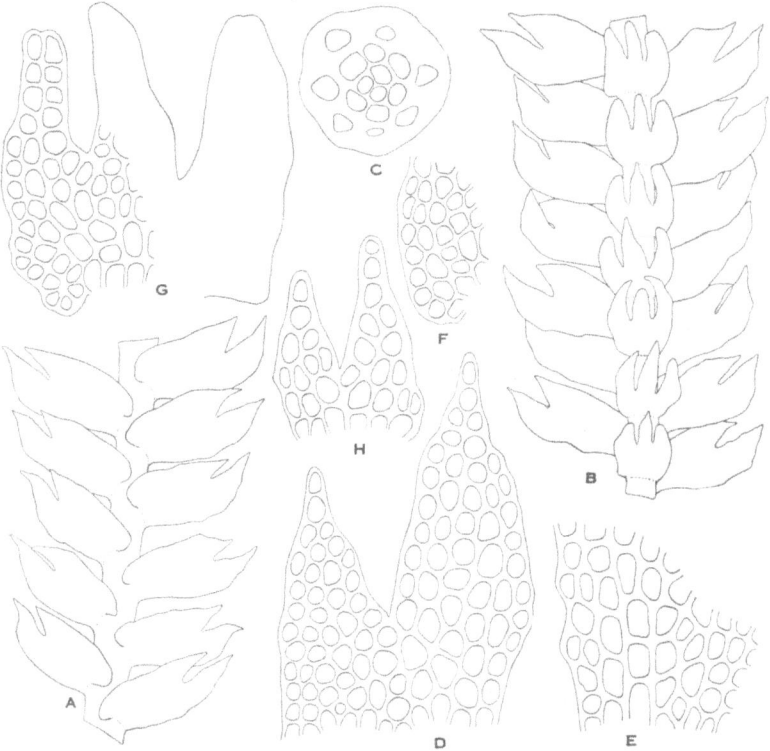

FIG. 13. *Acromastigum echinatiforme* (De Not.) Evans. *A*. Part of plant, dorsal view, × 50. *B*. Part of plant, ventral view, × 50. *C*. Cross-section of branch, × 225. *D*. Apical part of leaf, × 225. *E*. Cells from base of another leaf, ventral side, × 225. *F*. Dorsal base of leaf shown in *D*, × 225. *G*. Underleaf, × 225. *H*. Leaf of flagelliform branch, × 225. The figures were drawn from the specimens collected by BARTLETT in Sumatra.

tigum, in herbarium specimens, varies from pale to dark brown, and the plants are 1—2 cm. in length, with the dichotomies 2—4 mm. apart. The ordinary leafy branches are 0.08—0.11 mm. in width and 0.07—0.09 mm. in thickness. The cortical cells (Fig. 13, *C*) are about 30 μ in tangential width by about 25 μ in radial width, and the medullary cells average about 10 μ in diameter. The walls of the cortical cells are so strongly thickened that the cavities are only a little larger than those of the thin-walled medullary cells. The outermost walls, in fact, may attain a thickness of 10 μ.

The leaves are approximate to loosely imbricated (Fig. 13, *A*, *B*) and spread at an angle of about 60 degrees. In some cases they are more or less deflexed, in others plane or nearly so. They vary in outline from ovate-rectangular to obovate-rectangular and measure (on well-developed plants) 0.3—0.4 mm. in length by 0.12—0.18 mm. in width. At the dorsal base the leaves extend beyond the middle of the axis and develop a distinct rounded to subauriculate expansion (Fig. 13, *A*, *F*). Beyond the base the dorsal margin extends as a straight or slightly convex line to the apex of the dorsal division. The ventral margin, in most cases, is slightly convex throughout its entire length; on some of the leaves, however, it is straight or slightly concave toward the base and also along the ventral division. The oblique sinus, which is about one-third the length of the leaf, varies from acute to rounded. On some leaves the sinus is so narrow that the divisions lie almost in contact; on other leaves the sinus is wider, so that the divisions diverge at an angle not exceeding 30 degrees. The divisions are straight or nearly so and extend obliquely outward. Their apices are acute or short-acuminate and are tipped with a single cell or, more commonly, with a row of two cells (Fig. 13, *D*). The ventral division is longer and one or two cells wider than the dorsal division, which is three to five cells wide at the base. The leaf-margin is subentire throughout but presents a vaguely sinuous outline, owing to the fact that occasional cells project slightly in the form of very low crenulations (Fig. 13, *D*).

The cells in the ventral part of the leaf are slightly larger then those in the dorsal part; the difference in size is especially apparent toward the base but can be demonstrated also in the divisions (Fig. 13, *D*). The vaguely defined vitta (Fig. 13, *E*), which is two or three cells wide at the base, is separated from the ventral margin by two or three rows of cells and from the dorsal margin by five to seven rows. The cells of the vitta are 20—30 μ long and about 15 μ wide; those between the vitta and the dorsal margin average about 12 μ in diameter, those in the dorsal division measure about 14×12 μ, and those in the ventral division about 17×15 μ. The walls separating the cell-cavities appear, for the most part, uniformly thickened, but the angles of the cavities are rounded, and vague indications of trigones can occasionally be seen. The superficial walls are plane, and the cuticle minutely verruculose; the verruculae, however, are difficult to distinguish, except perhaps along the margin, where they appear in profile view.

The underleaves are distant to subimbricated (Fig. 13, B). They are somewhat convex, when seen from below, and lie subparallel with the axis, without being actually appressed to it. They are subquadrate in general outline and measure, on well-developed plants, 0.15—0.2 mm. in length. The narrow sinuses extend to the middle or a little beyond, and the divisions are subparallel or slightly convergent. These divisions, which are subequal in many cases, are three or four cells wide at the base and taper to truncate or subacute apices. The truncate divisions, which are in the majority, are tipped with two cells side by side (Fig. 13, G); between these cells the vestige of a slime-papilla can usually be detected, although there is little or no apical identation. The subacute divisions are tipped with a single cell or, more rarely with a row of two cells. At the base of the underleaf a minute rounded or subauriculate expansion (Fig. 13, G) is present on each side, and, in rare instances, a blunt tooth is developed a little higher up.

The flagelliform branches measure 0.06—0.09 mm. in diameter and are thus nearly or quite as large as the ordinary vegetative axes. Their ovate leaves (Fig. 13, H) are approximate but scarcely imbricated, and the more robust examples measure as much as 0.18 × 0.09 mm. In most cases, however, the leaves are considerably smaller. The acute sinus extends to or beyond the middle, and the pointed divisions are usually tipped with a row of two or three cells. At the base, on one or both sides, a minute expansion may be present, similar to the expansions at the base of the underleaves.

Most of the specimens of *A. echinatiforme*, which the writer has examined, are sterile, but the Amboina material includes a few plants of each sex. The male branches are more or less curved and, in the few examples seen, bear three to six pairs of imbricated, monandrous bracts. The latter are inflated and complicate-bifid about one third (Fig. 14, A, B), with a rounded and arched keel and acute or acuminate divisions, tipped with a single long cell or with a row of two cells (Fig. 14, C). Explanate bracts are broadly ovate and measure 0.3—0.35 mm. in length by 0.25—0.3 mm. in width. The ovate bracteoles are smaller than the bracts, and the example figured (Fig. 14, D) measures 0.25 × 0.15 mm. The bracteoles are slightly convex and are bifid about one-half with acute or rounded divisions. The margins of both bracts and bracteoles are entire; the walls of the cells are uniformly thickened, without distinct trigones; and the cuticle is smooth.

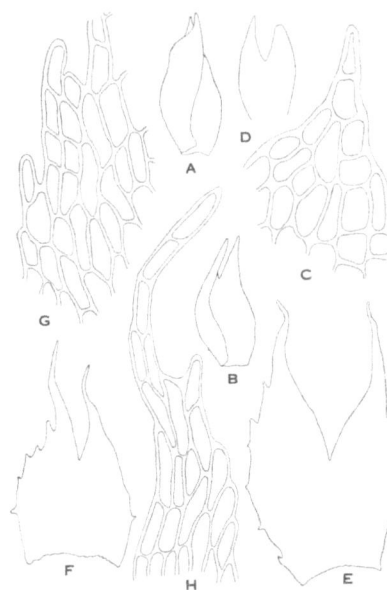

FIG. 14. *Acromastigum echinatiforme* (De Not.) Evans. *A*, *B*. Male bracts, × 50. *C*. Apex of division of male bract, × 225. *D*. Male bracteole, × 50. *E*. Perichaetial leaf of innermost series, × 40. *F*. Perichaetial leaf of second series, × 40. *G*. Cells from margin of perichaetial bract, innermost series, × 225. *H*. Apex of lacinia from mouth of perianth, × 225. The figures were drawn from the specimens collected by TEYSMANN on the island of Amboina.

DE NOTARIS has described the perichaetial leaves and perianth of *A. echinatiforme* and has illustrated them on his plate. Except in one or two unimportant details his statements and figures are in essential agreement with the observations made by the writer. As in other species of the genus the perichaetial leaves are in three or four series and increase rapidly in size toward the archegonia. Those of the innermost series are ovate (Fig. 14, *E*) and measure 1—1.2 mm. in length by 0.6.—0.75 mm. in width. They are bifid to the middle or a little beyond with an acute sinus and subulate, acuminate divisions, which are tipped, in the few examples seen, with a row of two long cells. The margins are sparingly and irregularly dentate or ciliolate-dentate (Fig. 14, G), although they are represented as entire in two of De Notaris' figures. The leaves of the next outer row (Fig. 14, F), which are 0.7—0.8 mm. in length and 0.04

—0.45 mm. in width, are similar to those of the innermost row.

The general features of the perianth are essentially like those of *A. anisostomum*. According to DE NOTARIS the mouth is surrounded by five subulate-capillary segments, each arising from a triangular base. This statement is supported by the only perianth available to the writer for dissection, which likewise shows five segments or laciniae around the mouth. It would be necessary, however, to examine a series of perianths before the number five could be accepted as a definite specific character. In all probability the number of laciniae is variable, just as it is in other species. Each lacinia is tipped with a row of from two to four long cells (Fig. 14, *H*), and the margin is either entire or vaguely and sparingly crenulate or denticulate. The cells of the perianth are 20—40 μ in length by about 10 μ in width, the cell-walls are uniformly thickened in a slight degree, and the cuticle is apparently smooth throughout.

DE NOTARIS very aptly compared the leaves of *A. echinatiforme* with the claws of crabs, and the resemblance is especially striking when the two sharp divisions closely approach each other. The rounded to subauriculate expansions at the dorsal base of the leaves represent one of the most distinctive features of the species. It is, unfortunately, easy to overlook these expansions, owing to the opacity of the axial organs, and no mention is made of them in the descriptions which have heretofore been published. The figures by DE NOTARIS likewise fail to show them and represent the dorsal margin of the leaf as a uniformly convex line, meeting the axis at an acute angle. The type-specimen of *Mastigobryum echinatiforme*, however, shows the expansions clearly, and the same thing is true of the other specimens listed by the writer. Similar but smaller expansions are found at the base of the underleaves.

The strongly thickened walls of the cortical cells and of the leaf-cells, together with their small size, indicate that *A. echinatiforme* is another xerophytic species. Since the outer walls of the cortical cells may be as much as 10 μ in thickness, these walls may occupy about one-fifth the thickness of the entire branch. This high ratio is duplicated, however, by the branches of all the other *Inaequilatera* which have so far been considered. The long and narrow divisions of the underleaves will at once distinguish *A. echinatiforme* from these species, the divisions of which are short and usually broad or even, in the case of *A. tenax*, more or less obsolete. The uniformly thickened walls of the leaf-cells, without trigones or pits, will further distinguish *A. echinatiforme* from *A. anisostomum*, the leaf-cells of which, at least in the ventral part of the leaf, show both trigones and pits.

12. **Acromastigum Brotheri** (Steph.) comb. nov.

Mastigobryum Brotheri Steph. Spec. Hepat. 3: 536. 1909.

The original specimens of the present species came from the herbarium of V. F. BROTHERUS and were collected by EVERETT on Mt. Poe, Borneo. According to the data given in STEPHANI's Icones, where the plant is figured, these specimens bear the number „59." The same number is represented in the MITTEN Herbarium by a packet upon which „*Bazzania Brotheri* Steph. n. sp." is written. This packet may therefore be considered authentic. Several other Bornean specimens in the MITTEN Herbarium, although deviating slightly from No. 59, could hardly be separated from it specificially. These specimens also were collected by EVERETT. An additional specimen has been received from Professor HERZOG. The species is known only from Borneo, and the various specimens examined by the writer may be recorded as follows:

B o r n e o: base of Mt. Poe, Sarawak, 800 ft. alt., 1892, A. H. EVERETT 59 (H., N. Y.), type of *Mastigobryum Brotheri* Steph.; Mt. Matang, Sarawak, no date, A. H. EVERETT (N. Y.); Mt. Bangok, Sarawak, no date, A. H. EVERETT (N. Y.); Kapuas River, Dutch Borneo, 1923, collector not named (Herz., Y.), specimens sent by R. WEGNER to T. HERZOG.

Although no illustrations of *Acromastigum Brotheri* have been published, the species is figured in STEPHANI's Icones and also in a series of original drawings by Professor SCHIFFNER, which have kindly been placed at the writer's disposal. These unpublished figures were all drawn from EVERETT's Mt. Poe material, and the same thing is true of Fig. 15 in the present paper. Fig. 16, however, which illustrates different forms of the species, was drawn from EVERETT's Mt. Matang material.

The specimens studied are all mixed with other bryophytes and grew scattered or in loose tufts or mats. The plants are yellowish brown or greenish brown in dried condition, and the living parts were evidently 0.5—1 cm. long. The dichotomies occur at intervals of 1—2 mm. Well-developed branches are 0.15—0.18 mm. in width and 0.12—0.15 mm. in thickness. The cortical cells, as seen in cross-

FIGURE 15. *Acromastigum Brotheri* (Steph.) Evans. *A*. Part of plant, ventral view, × 50. *B*. Cross-section of branch, × 225. *C*. Leaf, × 50. *D*. Cells from base of same leaf, ventral side, × 225. *E*. Cells from dorsal side of leaf, × 225. *F*. Apical part of leaf, × 225. *G, H*. Underleaves, × 50. *I*. Lateral division of underleaf shown in *G*, × 50. *J*. Leaf of flagelliform branch, × 50. The figures were drawn from the type-material.

section (Fig. 15, *B*) are 40—60 μ in tangential width by 30—40 μ in radial width, and the medullary cells average about 23 μ in diameter. The cell-walls in most cases are yellowish, the bounding walls of the cortical cells have a thickness of 8—10 μ, and the much thinner walls of the medullary cells show conspicuous thickenings at the angles.

The leaves (Figs. 15, *A*; and 16, *A, B*) are loosely imbricated and spread obliquely to widely, but the angle between a leaf and the axis rarely reaches 90 degrees. The upper leaf-surface is convex, especially along the ventral side, and the ventral margin in some cases appears in profile-view, when a branch is examined from below. The leaves are unsymmetrically ovate or ovate-rectangular in outline (Figs. 15, C; and 16, C) and measure, on well-developed plants, 0.4—0.55 mm. in length by 0.24—0.27 mm. in width. The dorsal base is rounded, without forming a distinct expansion, and the leaves arch to the middle of the axis or even a little beyond. Beyond the base the dorsal margin extends, on most leaves, as an approximately straight line. On some leaves, however, the line is slightly convex throughout; in others, slightly concave in the outer part. The ventral margin is approximately straight to the level of the sinus and then curves or bends forward. The ventral division, therefore, points obliquely forward, whereas the dorsal division points obliquely outward. The acute and oblique sinus is about one-fourth the length of the leaf and varies considerably in width, so that the divisions may lie almost in contact or diverge at an angle up to 30 degrees. On most leaves the divisions are acute and tipped with a single cell or with a row of two cells (Fig. 15, F), but on some leaves the divisions are blunter (Fig. 16, D), and the ventral division may even be rounded at the apex (Fig. 16, E). The ventral division, which is five to seven cells wide at the base, is longer and a little wider than the dorsal division, which is four to six cells wide at the base.

The vaguely defined vitta (Fig. 15, D) is two to four cells wide at the base and is separated from the ventral margin by two to four rows of cells and from the dorsal by eight to eleven rows. The cells of the vitta are 20—50 μ in length by 15—20 μ in width, those near the dorsal margin (Fig. 15, E) average about 15 μ in diameter, and those at the base of the divisions measure about $22 \times 16 \mu$. The cell-walls are uniformly thickened, without evident trigones, and the cuticle is apparently smooth throughout.

The loosely imbricated underleaves (Figs. 15, A; and 16, A, B) are distinctly convex, when seen from below, but are not appressed to the axis. They broaden out slightly from the base and show a broadly subquadrate or suborbicular form, with bulging sides (Figs. 15, G, H; and 16, F), when dissected off and spread out flat. Well-developed underleaves measure 0.2—0.25 mm. in length by about 0.3 mm. in width. The narrow sinuses, which extend to the middle or a little beyond, are acute to lunulate at the bottom; and the divisions are either subparallel or slightly divergent. In the majority of cases they are four to eight cells wide at the base and taper somewhat to a truncate (Fig. 16, G) or retuse (Fig. 15, I) apex, two to four cells wide, with a more or less distinct tooth on each side. In some instances, however, a division may be acute, owing to the non-development of one of the apical teeth. On many of the underleaves the divisions are subequal in width, but on some one or both of the lateral divisions may be wider than the median division. The margins are entire, except for the occasional presence of one or two bulges or teeth on one or both sides.

The flagelliform branches are 0.09—0.12 mm. in diameter and bear scattered to contiguous scale-like leaves. The latter may attain a length of 0.15 mm. They are ovate (Fig. 15, J) in general outline and are bifid to about the middle with sharp or blunt divisions.

The male branches are strongly curved in the form of circular arcs and may even form complete circles. The imbricated bracts, which are in eight to ten pairs, are inflated and complicate-bifid about one-half, with a rounded, strongly arched keel. Explanate bracts are ovate and measure about 0.3 mm. in length by 0.25 mm. in width. The acute or short-acuminate divisions are tipped with a single long cell or with a row of two cells, and the margins are vaguely and minutely crenulate or denticulate from projecting cells. The bracteoles, which also are ovate, are convex from below and measure only 0.18 mm. in length by 0.1 mm. in width. They, too, are bifid about one-half with sharp divisions. The cells of the bracts and bracteoles. although more delicate than ordinary leaf-cells, have slightly thickened walls without distinct trigones. The cuticle is apparently smooth throughout.

As here understood *A. Brotheri* shows a good deal of diversity with respect to the apices of the leaf-divisions and of the underleaf-divisions. In the leaf-divisions, for example, the apices are typically acute, as shown by Figs. 15, A, and 16, A. In some of the specimens listed, however, the divisions may be obtuse (Fig. 16, D, at right), and the ventral divisions may even be rounded (Fig. 16, E). These deviations from the acute type, moreover, may be found on many or all of the leaves on an individual plant or branch. The apices of the underleaf-divisions, similarly, may be retuse, truncate, or acute; and here again a given type of apex may be associated with many, if not all, of the underleaves of an individual branch. In Fig. 15, *A*, for example, none of the underleaves show acute divisions, whereas in Fig. 16, *A*, the lateral divisions of all the underleaves are acute. Of

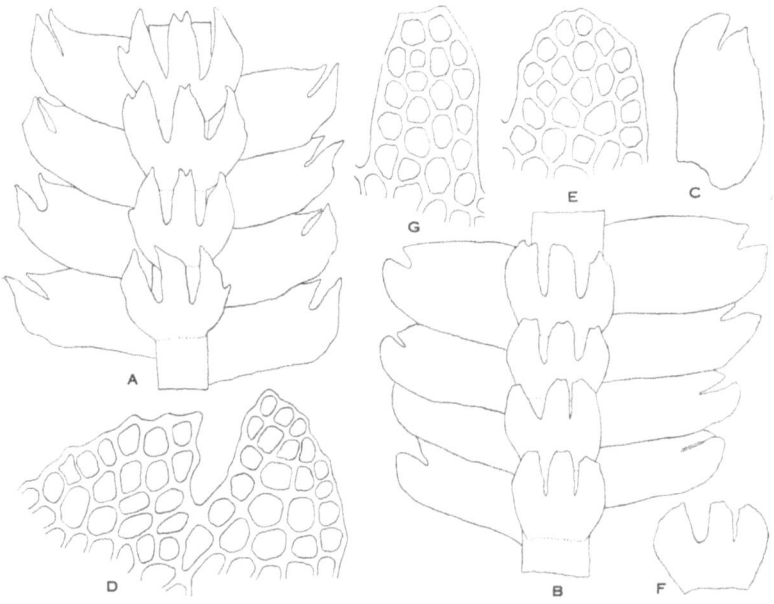

FIG. 16. *Acromastigum Brotheri* (Steph.) Evans. *A, B*. Parts of plants, ventral view, × 50. *C*. Leaf, × 50. *D*. Apical part of leaf, × 225. *E*. Ventral division of another leaf, × 225. *F*. Underleaf, × 50. *G*. Median division of same under-leaf, × 225. The figures were drawn from specimens collected by EVERETT on Mt. Matang, Borneo.

course the differences just cited are too slight and too inconstant to be anything more than specific variations.

The relationship between *A. Brotheri* and *A. echinatiforme* is very close. The plants are similar in color; they are both distinguished by bifid leaves in which the ventral division is a little longer and a little wider than the dorsal division; the leaf-margins of both are entire; the leaf-cells of both have uniformly thickened walls without distinct trigones or pits; and the underleaf-divisions in both are blunt in the majority of cases. Perhaps the most important distinction between the two is the presence of a definite expansion at the dorsal base of the leaves in *A. echinatiforme* and the absence of such an expansion in *A. Brotheri*. This difference, however, is accompanied by others. In *A. Brotheri*, for example, the cavities of the cortical axial cells are relatively large, although the walls are thickened; the broadest part of the leaf is below the middle; the leaf-divisions, although typically acute, may be obtuse or rounded; and the divisions of the underleaves in many cases, are more than four cells wide at the base and bidentate at the apex. In *A. echinatiforme*, on the other hand, the cavities of the cortical axial cells are relatively small, owing to the unusual thickness of the walls; the broadest part of the leaf, in many cases, is above the middle; the leaf-divisions are uniformly acute or short-acuminate; and the underleaf-divisions, which are not bidentate at the apex, are rarely, if ever, more than four cells wide at the base. The species, moreover, is smaller in all its parts than *A. Brotheri*, and this applies not only to the leaves and branches but also to the cells of which these organs are composed.

13. **Acromastigum exiguum** (Steph.) comb. nov.

Mastigobryum exiguum Steph. Hedwigia 25: 6. *pl. 2, f. 4—6.* 1885.

STEPHANI, in his original description of *A. exiguum*, states that the type-specimen was collected in southern Australia but does not mention the collector's name. In his S p e c i e s   H e p a t i c a r u m however, he is more explicit (39, p. 539) and states that the specimen was collected in Victoria by BÄUERLEN. The writer has seen a part of this specimen and also a second Australian specimen in the MITTEN Herbarium; they may be recorded as follows:

A u s t r a l i a : Victoria, without date, W. BÄUERLEN (H.), type of *Mastigobryum exiguum* Steph. from the STEPHANI collection; Sydney, New South Wales, A. COLLIE (N. Y.). No other specimens are known at the present time.

The plants are dull green, with little or no pigmentation of the cell-walls; the living portions are about 5 mm. long; and the dichotomies are 1—3 mm. apart. The main axes, when well developed, measure 0.12—0.15 mm. in width and 0.1—0.12 mm. in thickness. The cortical cells (Fig. 17, *A*) are 40—60 μ in tangential width by

30—40 μ in radial width, and the medullary cells average about 20 μ in diameter. The walls of the cortical cells are thickened and those on the outside may be as much as 10—12 μ thick; the walls of the medullary cells, however, are thin, except for the thickenings at the angles.

The imbricated leaves, which spread at an angle of about 60 degrees, are convex, when seen from above, and the apical portion may be so strongly deflexed that the divisions point toward the axis.

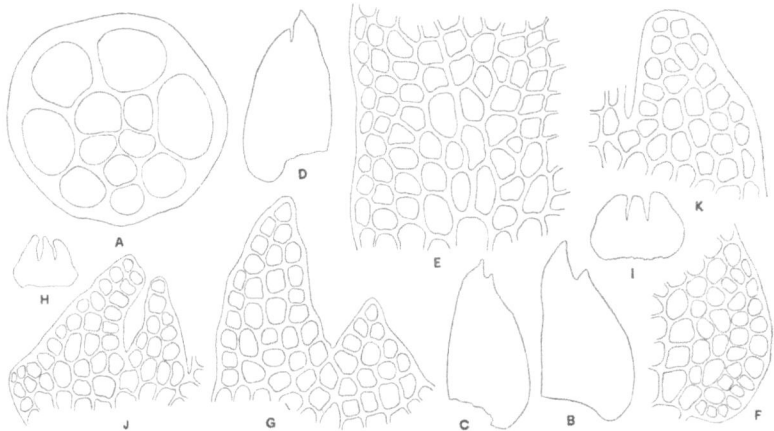

FIG. 17. *Acromastigum exiguum* (Steph.) Evans. *A*. Cross-section of branch, × 225. *B-D*. Leaves, × 50. *E*. Cells from base of leaf *B*, ventral side, × 225. *F*. Dorsal base of same leaf, × 225. *G*. Apex of same leaf, × 225. *H, I*. Under-leaves, × 50. *J*. Median and lateral divisions of underleaf *H*, × 225. *K*. Lateral division of underleaf *I*, × 225. The figures were drawn from specimens collected by COLLIE at Sydney, Australia.

In many cases the ventral margin appears in profile view toward the base, when a plant is examined from below, but this is not necessarily the case. When the leaves are dissected off and spread out flat (Fig. 17, *B—D*) they show an unsymmetrically ovate form but are not falcate. Well-developed examples measure 0.35—0.4 mm. in length by 0.18—0.25 mm. in width. They are rounded at the dorsal base (Fig. 17, *F*), beyond which the dorsal margin extends as a distinctly convex line to, or almost to, the apex of the dorsal division. The ventral margin is straight or slightly convex to the middle or beyond, but the outer part, in most cases, curves or bends gently forward. The sinus is acute to rounded, and the bottom is at one-ninth to one-fourth the distance from the tip of the ventral lobe to the base of the leaf. The divisions may be almost in contact but are usually divergent, and the angle between them may be as much as 60 degrees. On well-developed leaves both divisions are triangular and acute (Fig. 17, *G*) and are tipped with a single cell or, more rarely, with a row of two cells. The dorsal division is invariably smaller than the ventral and may be greatly reduced in size or even obsolete. In a series of leaves with distinct dorsal divisions, the latter were two to four cells long and two or three cells wide at the base, whereas the ventral divisions were six to eight cells long and three to five cells wide at the base. Leaves with reduced or obsolete dorsal divisions present the appearance of being undivided and acute, with or without a tooth on the dorsal margin close to the apex. The margins of the leaves are entire throughout.

The vitta is unusually indistinct, although an irregular group of elongate cells can be demonstrated in the basal part of the leaf close to the ventral margin (Fig. 17, *E*). If this group of cells is interpreted as the vitta, there are two or three rows of cells between the vitta and the ventral margin and nine or ten rows between the vitta and the dorsal margin. The cells of the vitta, which are not always in definite longitudinal rows, are mostly 25—30 μ long and about 20 μ wide; the cells in the dorsal part of the leaf toward the base average about 10 μ in diameter, and those in the apical part about 12 μ. The cells throughout the leaf have thickened walls, and the thickening usually appears uniform. In places, however, evidences of trigones may be apparent, although the pits between the trigones have been filled up by deposits of cell-wall substance. The walls in the vitta and vicinity often attain a thickness of 6 μ, but those in the dorsal part of the leaf are only 2 μ thick. The cuticle, in some cases, shows minute and indistinct verruculae but usually appears smooth.

The underleaves (Fig. 17, *H*, *I*), which are slightly convex when seen from below, are approximate to loosely imbricate and spread considerably beyond the axis on both sides. They are broadly orbicular in general form and measure, when well developed, 0.12—0.15 mm. in length by 0.17—0.22 mm. in width. The sides, which are distinctly convex, often have a rounded dilatation at the base and occasionally develop a blunt lobe-like protuberance higher up. The narrow acute sinuses extend to the middle or a little beyond, and the lateral divisions tend to converge toward the median division. The divisions have, in general, a triangular form, but the apex of the triangle is obtusely pointed or truncate, in most cases, and is formed by a single cell or by two cells side by side (Fig. 17, *J*, at left, *K*). Occasionally, however, a division is more acute and tipped with a row of two cells (Fig. 17, *J*, at right). Although the divisions may be subequal, the lateral divisions are usually broader than the median. In a series of underleaves examined the lateral divisions were four or five cells wide at the base, whereas the median divisions were three or four cells wide.

The ovate scale-like leaves of the flagelliform branches are distant to imbricated and measure, when well developed, about 0.1 mm. in length by 0.08 mm. in width. They are bifid to about the middle with an acute sinus and acute or truncate divisions. The sexual branches are still unknown.

STEPHANI, in his original description of the present species, emphasizes the undivided leaves as a significant feature, although he admits that an accessory tooth is present, in rare instances, below the apex. He brings out this feature clearly in two of his figures, one of which represents a leafy branch and the other three detached leaves. On the branch figured all the leaves are represented as undivided and entire, and two of the detached leaves are represented in the same way; the third detached leaf, however, shows a tooth on the dorsal margin just below the apex. In his revised account of the species (39, p. 538) he describes the leaves definitely as acute or apiculate and makes no mention of accessory teeth. The figures in his unpublished Icones were obviously drawn from the same preparations as his published figures but differ from them in a few details. The branch figured, for example, shows one leaf with a distinct tooth near the apex and another leaf with a vague subapical projection, but the detached leaves are all represented as undivided. The descriptions and figures cited indicate beyond a doubt that STEPHANI regarded *A. exiguum* as a species with normally undivided leaves.

Unfortunately the type-specimen does not support this idea very convincingly. Although some of the leaves are undivided, the majority are bidenticulate, bidentate, or shortly bifid. On one of the branches examined eight consecutive leaves on one side were bidenticulate, although the dorsal tooth in some cases was nothing more than a low rounded projection. On leaves with shortly bifid apices the dorsal divisions, although not exceeding two cells in length,were acute and distinct. It would appear, therefore, that *A. exiguum* is really a species with bifid leaves, and that the undivided leaves represent the exceptional condition. The trifid underleaves and the bifid leaves of the flagelliform branches support this idea. In species with characteristically undivided leaves, such as *A. bancanum*, the underleaves and the leaves of the flagelliform branches are likewise undivided.

The specimens collected by COLLIE have served for the writer's figures. These specimens, although obviously referable to *A. exiguum*, illustrate the bifid character of the leaves more clearly than the type-specimen. Every leaf examined, in fact, has shown some indication of a dorsal tooth or division. On some of the leaves the tooth consisted of a rounded bulge or a single projecting cell, but on better developed leaves a distinct dorsal division was present and this attained, in some instances, a length of three or four cells.

There is little danger of confusing *A. exiguum* with any of the other species of *Acromastigum*. The small dorsal divisions of the leaves, which exhibit so marked a variation in size, represent a very distinctive feature, especially since these small divisions are associated with definitely trifid underleaves. In the section *Exilia*, which is distinguished by minute or obsolete dorsal divisions, the underleaves are undivided; and in all the other species of the *Inaequilatera* with definitely trifid underleaves, the dorsal divisions are distinct and, in certain cases, even exceed the ventral divisions in size. STEPHANI compared *A. exiguum* with *A. anisostomum*, in which the leaves may be relatively small in comparison with the diameter of the axial organs. This species, however, is much more robust than *A. exiguum* and is further distinguished by a marked pigmentation of the cell-walls. The leaf-cells, too, are larger than those of *A. exiguum* and show distinct pits in the vitta and vicinity.

14. **Acromastigum Colensoanum** (Mitt.) Evans

Mastigobryum Colensoanum Mitt. in Hooker, Bot. Antarctic Voy. 2: 147. *pl. 100, f. 3*, 1855.
Mastigobryum divaricatum var. β. *Muellerianum* Gottsche, Linnaea 28: 556. 1857.

Mastigobryum minutulum Colenso, Trans. New Zealand Inst. 19:
    288. 1887.
Bazzania Colensoi Rodway, Papers & Proc. Roy. Soc. Tasmania
    1916: 75. 1916.
Acromastigum Colensoanum Evans in Reimers, Hedwigia 73: 142.
    1933.

The species was based on specimens collected by COLENSO in New
Zealand but was soon reported also from Australia and Tasmania.
The following specimens have been examined by the writer:
    A u s t r a l i a : Victoria, without date, F. VON MÜLLER (B.), type
of *Mastigobryum divaricatum* var. *Muellerianum* Gottsche; source of
the Yarra Yarra River, Victoria, without date, F. VON MÜLLER
(N. Y.); Blue Mountains, New South Wales, without date, A. COLLIE
(N. Y.). Reported also by STEPHANI (39, p. 539) from New South
Wales, on the basis of specimens collected by WATTS and WHITE-
LEGGE.
    N e w   Z e a l a n d : Tararua, North Island, without date, W.
COLENSO (N. Y.), type of *Mastigobryum Colensoanum* Mitt.; vicinity
of Norrewood, North Island, without date, W. COLENSO 1403 (Y.),
type of *Mastigobryum minutulum* Colenso; near Rotorua, North
Island, 1931 (No. H 433), and 1932, K. W. ALLISON (Hodg., Y.); Lake
Waikaremoana, North Island, 1932, E. A. HODGSON 1931 (Hodg.,
Y.); Petane, North Island, without date, A. HAMILTON 4852 (P., Y.);
Marlborough, South Island, without date, J. H. MCMAHON (Hodg.).
Reported also by STEPHANI (39, p. 539) from New Zealand, on the
basis of specimens collected by KIRK.
    T a s m a n i a : Brown River, without date, A. OLDFIELD (N. Y.),
cited and figured by BASTOW (2, p. 255, *pl. 26*); Mt. Wellington, 1897,
W. A. WEYMOUTH 1049 (H.), cited by STEPHANI (39, p. 539). ROD-
WAY (27, p. 75) reports the species also from Trowutta and the
Tasman Peninsula, without giving the names of the collectors.
    The plants grow on rotten logs or on the trunks of tree-ferns and
usually form depressed mats. They are delicate in texture and the
cell-walls are unpigmented throughout. The color therefore varies
from pale to dark green, fading to a yellowish hue with age. The
living portions are, for the most part, 0.5—1 cm. long, and the
dichotomies arise at intervals of 2—5 mm. The axes of the ordinary

vegetative branches are 0.15—0.3 mm. in width and 0.13—0.25 mm. in thickness, thus showing a rather wide variation in size. This is associated with a variation in the number of rows of cortical cells. The larger axes, for example, show the unusual peculiarity of having more than seven rows, and the axis figured (Fig. 18, *B*) showed fourteen to sixteen rows. The smaller axes, however, have seven rows (Fig. 18, *C*, *D*), as in all the other species of *Acromastigum* except *A. integrifolium*. The cortical cells are 40—75 μ in tangential width by 35—55 μ in radial width, and the medullary cells average about 30 μ in diameter. The bounding walls of the cortical cells attain a thickness of 6—8 μ, but the walls of the medullary cells are thin, except for the local thickenings at the angles.

The leaves (Fig. 18, *A*), which are plane or slightly convex, are distant to loosely imbricated and spread obliquely to widely, in some cases forming an angle of slightly more than 90 degrees with the axis. Well-developed leaves are 0.9—1 mm. in length by 0.4—0.6 mm. in width, but the leaves on more delicate plants or branches may be only 0.6 mm. in length by 0.3 mm. in width. The leaves are unsymmetrically ovate-rectangular in outline, and the larger examples (Fig. 18, *E*, *G*) are distinctly rounded at the base, arching to the middle of the axis or slightly beyond. The smaller leaves, however, although outwardly curved at the base, are scarcely rounded. Beyond the base the dorsal margin extends as a straight or slightly convex line to the apex of the dorsal division, which usually points outward. The ventral margin is subparallel with the dorsal and extends as an almost straight line to about the middle, beyond which it may be slightly convex. The acute to obtuse sinus is about one-third the length of the leaf and separates the divisions by an angle of 30—45 degrees. These divisions are narrowly to broadly triangular and taper to an acute apex, which is tipped with a single cell (Fig. 18, *H*) or with a row of two cells (Fig. 18, *I*). The divisions are much alike, although in most of the leaves the ventral division is a trifle narrower and a trifle longer than the dorsal. Cases are not infrequent, however, in which the ventral division is slightly shorter than the dorsal, and several such cases are shown in Fig. 18, *A*. In a series of large leaves examined the ventral divisions were six to eight cells wide at the base and the dorsal nine to twelve cells wide, but in a series of smaller leaves the ventral divisions were only three or four

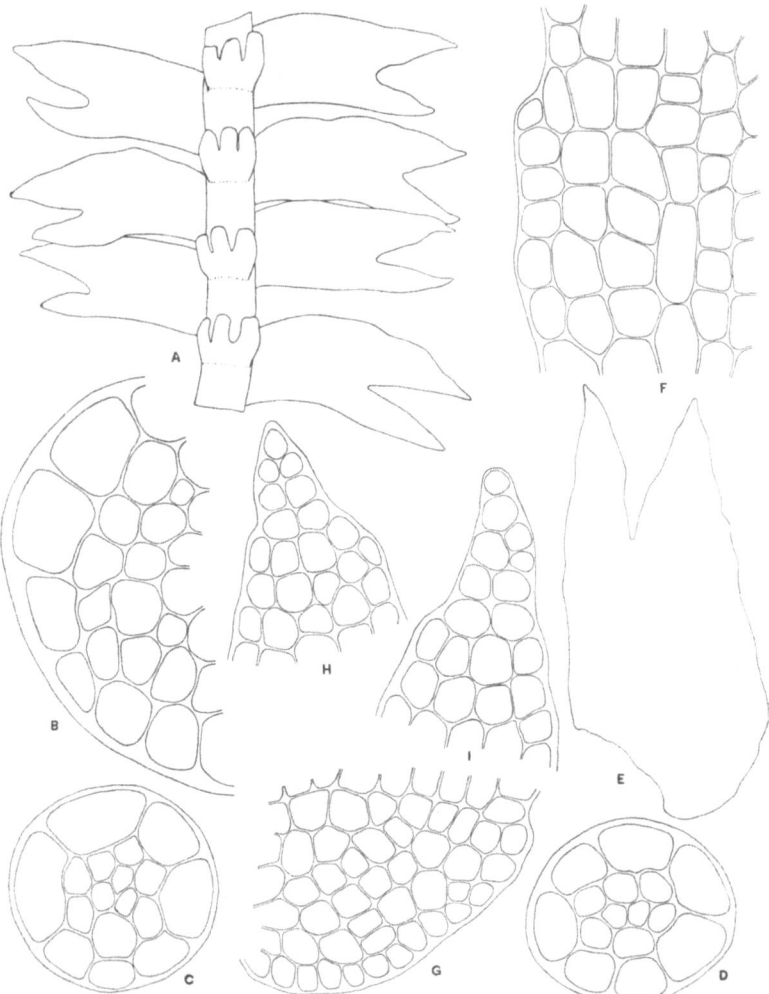

FIG. 18. *Acromastigum Colensoanum* (Mitt.) Evans. *A*. Part of plant, ventral view, × 50. *B*. Cross-section of robust branch, × 225. *C, D*. Cross-sections of slender branches, × 225. *E*. Leaf, × 50. *F*. Cells from base of same leaf, ventral side, × 225. *G*. Dorsal base of same leaf, × 225. *H*. Apex of dorsal division of same leaf, × 225. *I*. Apex of ventral division of same leaf, × 225. *A* and *D* were drawn from the type-material of *Mastigobryum minutulum*; the remaining figures, from the specimens collected by OLDFIELD at Brown River, Tasmania.

cells wide and the dorsal only six to eight cells wide. The leaf-margins are entire throughout, except for the presence of vague and irregular sinuations.

The vitta is very vaguely defined even toward the base. On well-developed leaves it consists of a short band two to four cells wide, separated from the ventral margin by one to three rows of cells (Fig. 18, *F*) and from the dorsal margin by fifteen to twenty rows. The cells of the vitta, as a rule, measure 20—25 μ in width by 25—50 μ in length, but a cell 60 μ long can occasionally be demonstrated. Toward the dorsal margin the cells average about 20 μ in diameter, in the divisions about 25 μ. The contrast in size, therefore, is not very pronounced. The walls between the cells are thin throughout, except for minute trigones with concave sides; but the walls in the region of the vitta may be a trifle thicker than the others. In many cases two trigones will coalesce and form a four-sided figure. The bounding walls of the cells are always distinctly thickened, as shown by the walls along the margin; and the cuticle is either smooth or minutely and very indistinctly verruculose. In poorly developed leaves, which may be only eight to ten cells wide at the base, the vitta is separated from the ventral margin by only one or two rows of cells and from the dorsal margin by only five or six rows.

The underleaves are distant and appressed to the axis. They are quadrate or quadrate-rectangular in general outline, with straight or slightly bulging sides, and measure, in well-developed examples, 0.25—0.3 mm. in length by 0.3—0.35 mm. in width. The sinuses, which are one-third to one-half the length of the underleaf, may be very narrow and acute or somewhat broader and obtuse to rounded. The divisions lie subparallel (Fig. 19, A—C) and have parallel or somewhat convergent sides and slightly retuse, truncate, or bluntly pointed apices. On well-developed underleaves the divisions are three to five cells wide at the base (Fig. 19, *D*) and one to four cells wide at the apex; on poorly developed underleaves, however, they may be only two cells wide at the base. The cells and margins of the underleaves are essentially like those of the leaves.

The flagelliform branches, even on poorly developed plants, have a diameter of about 0.12 mm. Their scale-like leaves, which may attain a length of 0.1 mm., are ovate to oblong and bifid about one-half, with a narrow sinus and truncate to acute divisions.

According to RODWAY (27, p. 75) the perianth of *A. Colensoanum*, which is 3.5 mm. long, is fusiform and tapers toward the apex. Except for this statement nothing is said about the female inflorescence in the published descriptions of the species, and the male inflorescence is apparently still unknown. Most of the specimens studied by the writer are sterile. OLDFIELD's Tasmanian material, however, contains a few female branches with very immature perianths, and one of ALLISON's New Zealand specimens shows a few mature perianths.

The perichaetial leaves, as in other species of *Acromastigum*, are in three or four series. Those of the innermost series (Fig. 19, *E*) are ovate from a broad base and measure 1.1.—1.25 mm. in length by 0.65—0.75 mm. inwidth. A narrow sinus, extending nearly or quite to the middle, divides each leaf into two subulate, acuminate divisions, tipped with a single cell or with a row of two or three cells. The margins are sparingly and irregularly crenulate or denticulate and may show, on one or both sides, one or two larger sharp-pointed teeth or laciniae. The cells tend to be longer than broad and are thin-walled throughout (Fig. 19, *F*), although minute trigones can sometimes be demonstrated. The cuticle is faintly striolate-verruculose. The leaves of the second series from the inside are 0.9—1 mm. in length by 0.5—0.6 mm. in width, and their marginal projections are less pronounced than in the leaves of the innermost series.

One of the perianths examined by the writer had a length of 6 mm. and a diameter of 0.9 mm.; in other respects it agreed with RODWAY's description. The perianths show the usual tricarinate condition in the upper part and are divided at the mouth into about twelve laciniae, which vary considerably in both length and width (Fig. 19, *G*). The cells at the apices of the laciniae are slender and cilium-like, and a row of three or four such cells may be present. The sides of the larger laciniae are sparingly and irregularly denticulate or ciliolate, and the majority of the teeth and cilia are unicellular. The cells of the perianth, throughout most of its extent, have an average width of 20 μ, but their length varies from 50 μ to 120 μ. The walls are thin, except for minute trigones, and the cuticle is either smooth or faintly striolate.

The most remarkable feature of *A. Colensoanum* is the arrangement of the cortical cells on robust axes in more than seven longi-

tudinal rows. The only other species in which such an arrangement occurs is *A. integrifolium*, which is of course at once distinguished by its
transversely attached and undivided leaves and underleaves. The
arrangement in more than seven rows, unfortunately, is not constant
in *A. Colensoanum*. The more delicate branches, even on robust
individuals, may show only seven rows, and the same number of

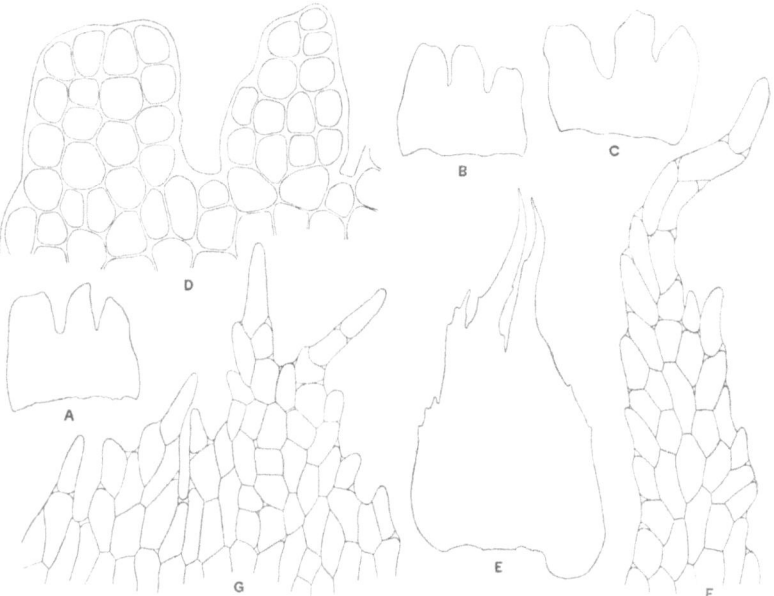

FIG. 19. *Acromastigum Colensoanum* (Mitt.) Evans. *A-C*. Underleaves, × 50.
D. Median and lateral divisions of underleaf *A*, × 225. *E*. Perichaetial leaf of
innermost series, × 40. *F*. Apex of division of perichaetial leaf, × 225.
G. Laciniae from mouth of perianth, × 225. The figures were drawn from the
specimens collected by OLDFIELD at Brown River, Tasmania.

rows may be present on all the branches of poorly developed
material. COLENSO's *Mastigobryum minutulum* was based on material
of this character and is distinguished with difficulty from the
following species. The writer has followed STEPHANI (37, p. 275),
however, at least provisionally, in including *M. minutulum* among
the synonyms of *A. Colensoanum*. It should be noted that the flagelliform branches, even when springing from robust axes, show only
six rows of cortical cells.

15. **Acromastigum divaricatum** (Nees) Evans

Jungermannia divaricata Nees, Hepat. Javan. 60. 1830. Not *J. divaricata* Sm. Engl. Bot. *pl. 719.* 1800.
Mastigobryum divaricatum Nees in G. L. & N. Syn. Hepat. 219. 1845.
Bazzania divaricata Trevis. Mem. Ist. Lomb. 13: 414. 1877.
Acromastigum divaricatum Evans in Reimers, Hedwigia 73: 142. 1933.

NEES VON ESENBECK based his *Jungermannia divaricata* on sterile material collected in Java by BLUME. This material has served for the illustrations published by LINDENBERG and GOTTSCHE (20, *pl. 5, f. 1—5*) and for a series of original drawings kindly sent by Professor SCHIFFNER. According to LINDENBERG and GOTTSCHE's f. 3 the underleaves may be trifid, bifid, or irregularly toothed or lobed at the apex; according to Professor SCHIFFNER's drawings they are definitely trifid, and an accompanying note states that the underleaves in LINDENBERG and GOTTSCHE's figure were incorrectly drawn. Through the kindness of Dr. REIMERS the writer has been able to examine a specimen from the BRAUN Herbarium, which probably represents a part of BLUME's original material. This specimen also has trifid underleaves and agrees in this important respect with the various other specimens listed below. Strange to say, STEPHANI's figures in the Icones represent the underleaves as predominantly bifid, and thus awaken the suspicion that they must have been drawn from poorly developed plants. The following specimens have been examined by the writer:

J a v a: without definite locality or collector's name (B.), specimen from the BRAUN Herbarium, probably representing a part of BLUME's original collection; Mts. Gedeh and Salak, without date, J. E. TEYSMANN (B., H., L.), cited by SANDE LACOSTE (28, p. 302).

M a l a c c a: Mt. Ophir, Johore, without collector's name or date (N. Y.), specimen from the GRIFFITH Herbarium.

P h i l i p p i n e s: Mt. Maquiling, province of Laguna, Luzon 1916, C. B. ROBINSON 9747, 17089, 17097, 17113, 17155 (N. Y., Y.), all distributed by the Bureau of Science, Manila; same locality, 1916, C. J. BAKER 7007 (Herz., Y.), cited by HERZOG (17, p.83).

S u m a t r a: Mt. Singalang, 1720—2130 m. alt., 1894, V. SCHIFF-

NER (V., Y.); Mt. Barisan, Expedition of 1878 (H., L.). Reported also by STEPHANI from Sumatra (39, p. 537), on the basis of specimens collected by De Vrieze.

SCHIFFNER (32, p. 153) reports the species also from Amboina, Australia, Tasmania, and the Straits of Magellan. His record for Amboina was based on specimens collected at Salhaetoe by KARSTEN in 1889 (30, p. 258), and his record for the Straits of Magellan on specimens collected by NAUMANN in 1876 (29, p. 18). The latter specimens, which are now in the Cryptogamic Herbarium of Harvard University, are sterile and very fragmentary. It is possible that they represent a small form of *A. Colensoanum*, but it would be unwise to base a definite record upon them. It is also possible that SCHIFFNER's records for Australia and Tasmania were based on specimens of *A. Colensoanum*, since the true *A. divaricatum* has not recently been collected outside the Indo-Malayan region.

The present species, as here understood, exhibits a rather wide range of variability, but the different forms intergrade into one another so gradually and so completely that any subdivision of the species seems inadvisable. The plants are delicate and grow in depressed mats, often mixed with other bryophytes. The color varies from a whitish green to a pale yellowish green, becoming pale brownish with age, but the cell-walls remain unpigmented throughout. The living parts are 1.5—2 cm. long; the dichotomies, which spread at an acute angle, occur at intervals of 2—8 mm.; and well-developed axes measure 0.15—0.17 mm. in width by 0.12—0.15 mm. in thickness, thus showing a slight dorsiventral compression. The tissues are so transparent that the seven rows of cortical cells and the contrast in size between cortical and medullary cells can easily be demonstrated in intact plants. The cortical cells, as seen in cross-section (Fig. 20, $\dot{B}$), have a tangential width of 40—70 µ and a radial width of 30—50 µ, whereas the medullary cells have an average diameter of about 25 µ. The bounding walls of the cortical cells are 4—8 µ thick, but the walls of the medullary cells are thin, except for minute thickenings at the angles. For comparison with an ordinary vegetative branch the cross-section of a flagelliform branch is figured (Fig. 20, M). This shows only six cortical cells and three medullary cells.

The leaves (Fig. 20, A) are distant to loosely imbricated and spread obliquely to widely, forming an angle of 60—90 degrees with the axes. Some of the leaves are more or less deflexed, others lie approximately in a single plane, and it is not unusual for a single plant to show both of these positions. Well-developed leaves (Figs. 20, *C—F*; and 21, *A, B*) are rectangular to ovate-rectangular in outline, with a slight asymmetry, and measure 0.45—0.8 mm. in length by 0.15—0.3 mm. in width. The dorsal base is slightly rounded (Fig. 21, *D*) but scarcely reaches beyond the middle of the axis. Beyond the base the dorsal margin extends as a straight or slightly concave line to the apex of the dorsal division, which in most leaves points obliquely forward. The ventral margin is approximately parallel with the dorsal throughout the greater part of its length. It extends as a nearly straight line almost or quite to the base of the ventral division, the lower margin of which may show a slight convexity or concavity. The ventral division diverges more or less from the dorsal and points obliquely outward rather than forward. Both divisions are acute and are tipped with a row of two cells (Figs. 20, *H, I*; and 21, *F*) or with a single cell (Fig. 21, *E*). The sinus between them is one-fourth to one-third the length of the leaf and varies from acute to rounded at the bottom. The divisions are less diverse than in most of the other species of *Acromastigum* with bifid leaves, but the ventral division is, in most cases, narrower and slightly longer than the dorsal. In a series of leaves examined the ventral divisions were two to four cells wide at the base and the dorsal three to seven cells wide. The leaf-margins are entire throughout.

The vitta, as in *A. Colensoanum*, is very indistinct. It consists of a short group of cells, one to three cells wide, at the base of the leaf (Figs. 20, *G*; and 21, *C*) and is separated from the ventral margin

Fig. 20. *Acromastigum divaricatum* (Nees) Evans. *A*. Part of plant, showing dorsiventral and flagelliform branches, ventral view, × 50. *B*. Cross-section of branch, × 225. *C-F*. Leaves, × 50. *G*. Cells near base of leaf, ventral side, × 225. *H, I*. Apices of leaves, × 225. *J*. Underleaf, × 50. *K, L*. Underleaves, × 225. *M*. Cross-section of flagelliform branch, × 225. *N*. Leaf of flagelliform branch, × 225. *A, C, D* and *J* were drawn from specimens collected by ROBINSON on Mt. Maquiling, Philippines, No. 17113; *B, G, H, K, M* and *N*, from specimens collected by ROBINSON in the same locality, No. 9147; and *E, F, I* and *L*, from specimens collected by SCHIFF-NER on Mt. Singalang, Sumatra, No. 377.

by one or two rows of cells and from the dorsal margin by five or six rows. The cells of the vitta are 25—35 μ in length by about 24 μ in width, the cells near the dorsal margin average about 15 μ in diameter, and the cells in the divisions average about 24 μ. The bounding

walls of the cells are thickened, as shown by the marginal walls, but the cells are otherwise thin-walled, except for small but distinct trigones, with concave or straight sides. Four-sided figures, each formed by the coalescence of two trigones are frequent. The cuticle is faintly striolate or striolate-verruculose.

The underleaves (Fig. 20, *A*), which are distant to approximate, are appressed to the stem and therefore slightly convex, when seen from below. They show a subquadrate outline (Figs. 20, *J*; and 21, *G*, *H*), with straight or slightly bulging sides, and measure 0.08—0.15 mm. in length. The divisions, in most cases, are two to four cells wide at the base (Fig. 20, *K*, *L*) and somewhat narrower at the apex, which varies from truncate to subacute. They are separated by acute to obtuse sinuses extending about half-way to the base. In most of the underleaves the divisions are unequal. When an apex is truncate it is tipped with two or (rarely) three cells side by side and shows little or no apical depression; when an apex is subacute it is tipped with a single cell or with a row of two cells. The margins of the underleaves are entire, as in the leaves, and the cells are essentially like the leaf-cells.

Flagelliform branches are abundantly produced in *A. divaricatum* (Fig. 20, *A*) and have a diameter of 0.08—0.12 mm. Their distant leaves are subquadrate (Fig. 20, *N*) and measure, in well-developed examples, 0.08—0.1 mm. in length. The divisions are truncate to subacute and are separated by a sinus about half as long as the leaf.

The sexual branches of *A. divaricatum* have not yet been described, and most of specimens examined by the writer are completely sterile. A few of the plants, however show male branches or female branches with unfertilized archegonia. The male branches arise in the axils of ordinary underleaves or on flagelliform branches. They are more or less curved and bear six to ten pairs of imbricated monandrous bracts. The latter are inflated and complicate-bifid about one-half, with a rounded, strongly arched keel and acute divisions, which are tipped with a row of two cells (Fig. 21, *J*) or with a single cell. Explanate bracts (Fig. 21, *I*) are broadly ovate and measure 0.25—0.3 mm. in length by 0.2—0.25 mm. in width. The bracteoles (Fig. 21, *K*) are smaller than the bracts and measure about 0.1 mm. in length; the majority are bifid with rounded or truncate divisions, but the bracteole at the base of the branch may be trifid. The margins of the bracts and bracteoles are entire; the cells are thin-walled and rarely show minute trigones; and the cuticle is finely and faintly striolate-verruculose.

In female branches with unfertilized archegonia it is difficult to determine whether the structures immediately surrounding the archegonia represent the apices of undeveloped perichaetial leaves or the laciniae at the mouth of a rudimentary perianth. They are here interpreted in the latter way. There is therefore a possibility that the next outer series of structures, which are here described as the leaves of the innermost series, may really represent the leaves of the second series outside the archegonia. The perichaetial leaves, as here interpreted, are in three or four series. Those of the innermost series (Fig. 21, L) are narrowly ovate and measure about 1.2 mm. in length by 0.4 mm. in width. They are bifid to about the middle with a narrow sinus and acuminate divisions tipped with a row of two to four cells. The margins are nearly entire in most cases, showing nothing more than vague and scattered crenulations or denticulations. Some of the leaves, however, bear one or two irregular teeth or lobe-like projections toward the base. The leaves of the next outer series (Fig. 21, M) are shorter and relatively broader than those of the innermost series, measuring 0.7—0.9 mm. in length by O.35—0.45 mm. in width. Their divisions taper more abruptly, being tipped in most cases with a single cell, and their marginal teeth tend to be more distinct. The cells of the perichaetial leaves, at least in the divisions, are (with occasional exceptions) longer than broad; the walls are thin but may show minute trigones; and the cuticle is distinctly striolate-verruculose.

The structures immediately surrounding the archegonia are in the form of long and slender laciniae, each tipped with a cilium consisting of a row of two to eight cells. The margins of the laciniae are, for the most part, entire or very sparingly crenulate or denticulate, but occasionally a more distinct tooth in the form of a short cilium is present. The cells, especially toward the base, are very thin-walled, but are otherwise much like the cells of the perichaetial bracts, as described above.

In the specimens from Java and Sumatra the leaves show a strong tendency to be deciduous, thus serving as organs of vegetative multiplication. The deciduous leaves are essentially like the other leaves and show no peculiarities associated with their deciduous habit. In the specimens from Malacca and the Philippines the leaves seem to be firmly attached throughout, and it might at first seem that the lack of the deciduous habit would be sufficient to separate

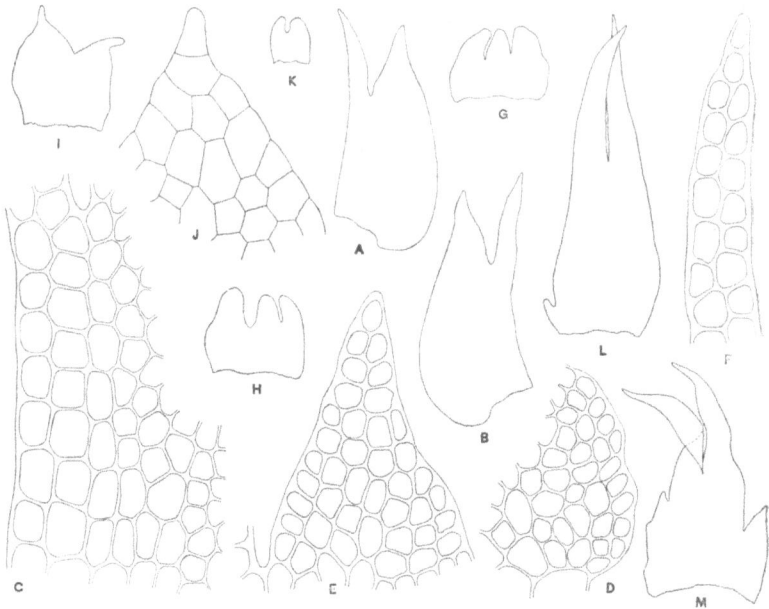

FIG. 21. *Acromastigum divaricatum* (Nees) Evans. *A*, *B*. Leaves, × 50. *C*. Cells from base of leaf *A*, ventral side, × 225. *D*. Cells from dorsal base of same leaf, × 225. *E*. Dorsal division of same leaf, × 225. *F*. Ventral division of leaf *B*, × 225. *G*, *H*. Underleaves, × 50. *I*. Male bract, × 50. *J*. Apex of division of male bract, × 225. *K*. Male bracteole, × 50. *K*. Perichaetial leaf of innermost series, × 40. *M*. Perichaetial leaf of second series, × 40. *A-H* were drawn from specimens in the GRIFFITH Herbarium collected on Mt. Ophir, Johore, Malacca; *I-K*, from specimens collected by ROBINSON on Mt. Maquiling, Philippines, No. 17089; and *L* and *M*, from specimens collected by SCHIFFNER on Mt. Singalang, Sumatra, No. 377.

these specimens specifically from the others. The writer, however, has been unable to detect any other differences of importance, and it therefore seems advisable to include all the specimens cited under a single species.

It should perhaps be noted in this connection that STEPHANI, in his unpublished Icones, figured some of the Philippine specimens listed above (Nos. 17089, 17097, 17155, 17158) as a distinct species, to which he gave a manuscript name. When the figures are compared with his figures of *A. divaricatum*, to which attention has already been called, it will be seen that an important difference in the under-leaves is indicated. In the manuscript species these are represented as trifid, whereas in *A. divaricatum* they are represented as bifid. Whether STEPHANI based his new species on this difference is not stated, but is perhaps probable, since no other important differences are brought out. This difference, however, is without significance because, as already shown, the underleaves of *A. divaricatum* are normally trifid. STEPHANI makes no allusion to the manuscript species in his Species Hepaticarum and therefore could have had but little faith in its validity.

The variability of *A. divaricatum* is shown primarily in the form and relative size of the leaf-divisions, in the width of the sinus, in the form and relative size of the underleaf-divisions, and in the thickness of the walls of the leaf-cells. The differences in the form and size of the leaf-divisions are largely differences in width. In some of the leaves, as shown by Fig. 20, *I*, the divisions are almost equal in width, and the ventral division (on the right) is distinguishable from the dorsal only through its slightly greater length and slightly larger cells. At the other extreme, as shown by Fig. 21, *F*, the ventral division is subulate and only two cells wide throughout most of its length, whereas the dorsal division (Fig. 21, *E*) is more broadly triangular and seven cells wide at the base. Most of the leaves, however, as shown by Fig. 20, *A*, occupy a position between these extremes. The slightly greater length of the ventral division represents an almost constant specific character. It is only in rare instances that the dorsal division is the longer of the two.

The figures will give some idea of the variability in the width of the sinus and in the form and size of the underleaf-divisions. In some of the leaves the sinus is even narrower than in any of the examples shown, and the divisions lie almost in contact with each other. The variability in the thickness of the walls of the leaf-cells is shown particularly by differences in the size and distinctness of the trigones. In the material from Sumatra (Fig. 20, *G—I*) these are well-developed and are separated from one another by definite thin areas. The specimens from Java and the Philippines are in close agreement with the Sumatra material, except that the trigones may be a trifle smaller. In the specimens from Malacca the trigones of some of the cells, as shown by Fig. 21, *C—E*, are less distinct, and the walls present the appearance of being uniformly thickened. Other cells, however, in the same material, show more distinct trigones and thus bridge over the gap between the cells figured and the cells in the Sumatra specimens.

Although *A. divaricatum* is closely related to *A. Colensoanum* there is no difficulty in telling them apart, if the specimens of the Australasian species are normally developed. Such specimens will always show robust branches in which more than seven longitudinal rows of cortical cells are present, whereas the number seven is apparently never exceeded in the internodes of *A. divaricatum*. The leaves and underleaves of *A. Colensoanum* moreover, particularly those borne on the robust branches, are distinctly larger than those of *A. divaricatum*. The difficulty of separating the species is increased, however, if the specimens of *A. Colensoanum* are poorly developed and show no branches of the robust type. In such specimens, which are illustrated by COLENSO's material of *Mastigobryum minutulum*, the axes agree with those of *A. divaricatum* in having only seven rows of cortical cells, and any differences in the size of the leaves and underleaves are negligible. Under such circumstances there is little to rely upon, as a differential character, except the greater opacity of the plants in *A. Colensoanum*. In *A. divaricatum* the branches and leaves are unusually transparent and the walls of the cortical cells stand out with glass-like distinctness.

16. **Acromastigum laetevirens** (Ångstr.) comb. nov.

Mastigobryum laetevirens Ångstr. in Stephani, Hedwigia 25: 133. *pl. 4, f. 4—6.* 1886.
Bazzania laetevirens Steph. Hedwigia 32: 206. 1893.

According to STEPHANI's original description the present species was given the manuscript name *Mastigobryum laetevirens* by VAN DER SANDE LACOSTE and was based on specimens collected by KRAUSE at Corral, Chile. In his Species Hepaticarum, however (39, p. 540), he accredits the name to ÅNGSTRÖM, instead of to VAN DER

SANDE LACOSTE, and leaves in doubt the name of the collector of the Corral specimens. The figures in his Icones were drawn from KRAUSE's specimens and agree in all essential respects with those in his original publication. It is evident, therefore, that the species was largely based on these specimens, even if they may not have been the actual plants originally named by ÅNGSTRÖM. The only specimen of *A. laetevirens* studied by the writer is a portion of KRAUSE's material in the GOTTSCHE Herbarium; this may be recorded as follows:

C h i l e: Corral, without date, K. KRAUSE (B.). STEPHANI reports the species also from two additional Chilean stations: Valdivia, HAHN (39, p. 540), and Hale Harbor, C. SKOTTSBERG (40, p. 60).

The specimen examined by the writer consists of a single plant 7 mm. in length, with dichotomies 2.5—3 mm. apart. The color is dingy green and the walls seem to be unpigmented throughout. The main axes are about 0.15 mm. in width, but no details can be given regarding the cortical and medullary cells, since the material is too scanty for sectioning or dissection. The account of the leaves and underleaves, also, is less satisfactory than might be desired.

The leaves are loosely to rather closely imbricate and spread at angles of 60 to 90 degrees. The upper surface is convex and the ventral and apical portions are more or less deflexed. The deflexed portion includes the ventral division and, in some cases, the dorsal division as well. The leaves are ovate and measure 0.5—0.6 mm. in length by 0.25—0.3 mm. in width. At the dorsal base a rounded dilatation is present. The divisions are sharp-pointed, and the apices are formed by a row of two cells or by a single cell. The dorsal division, which is triangular, is nine to ten cells wide at the base. The ventral division, which is subulate, is narrower than the dorsal, being only four or five cells wide at the base, and is usually shorter as well. The margin is irregularly and very vaguely crenulate from projecting cells.

An indistinct group of somewhat elongate cells can be demonstrated at the base of the leaf near the ventral margin. The cells of this group, which is hardly definite enough to be called a vitta, may attain a size of $25 \times 20 \mu$. The cells in the divisions average about 20μ in diameter and those in the dorsal part of the leaf near the base about 17μ. The cell-walls are thin, but small trigones with concave

sides are present, and occasionally two trigones coalesce. The superficial walls bulge slightly, and the cuticle is minutely and closely
verruculose or striolate-verruculose.

The broadly orbicular underlaeves, which are loosely imbricate and
convex when seen from below, are 0.2—0.25 mm. long and 0.25—
0.3 mm. wide. The narrow sinuses extend to about the middle, and
the broad subparallel divisions are rounded to truncate at the apex.
According to a series of underleaves examined the divisions are six
to eight cells wide at the base and three or four cells wide at the apex.
The leaves of the flagelliform branches, which are orbicular and
about 0.15 mm. in length, are bifid to the middle with broad divisions,
the apices of which vary from acute to truncate. The sexual branches
are still unknown.

STEPHANI emphasized the deflexed leaves of *A. laetevirens* in his
original account and stated that this feature would distinguish the
species from all the other *Inaequilatera*. Deflexed leaves, however,
are not peculiar to *A. laetevirens*. In *A. divaricatum*, for example,
some of the specimens cited have leaves that are more or less
deflexed, although the majority of the leaves are explanate; and the
same thing is true of a few other species. The deflexed leaves of *A.
laetevirens* would, nevertheless, represent an important specific
character, especially if the constancy of this feature could be
established.

In his published description STEPHANI tells us nothing about the
allies of *A. laetevirens*, but in his unpublished Icones he notes the
proximity of the species to *A. divaricatum* and *A. Colensoanum*. His
figure of an explanate leaf (*f. 5*) indicates that the proximity must
indeed be very close. Just as in the other two species the outline of
the leaf is ovate-rectangular, the dorsal base is rounded, the margin
is entire or nearly so, and the ventral division is narrower than the
dorsal. The figure shows a ventral division, however, which is not
only narrower than the dorsal but also distinctly shorter. In *A.
divaricatum* the ventral divisions are almost invariably longer than
the dorsal; in *A. Colensoanum* a similar relation prevails, although
leaves in which the ventral divisions are shorter than the dorsal are
not infrequent. If it could be demonstrated that the ventral divisions
in *A. laetevirens* are, in the majority of the leaves, shorter than the
dorsal, a specific distinction might be based on this feature, supple-

menting the distinction based on the deflexed leaves. Unfortunately the scanty material available at the present time makes such a demonstration impossible, and the claims of the species for recognition must therefore remain somewhat inconclusive.

17. **Acromastigum curtilobum** (Schiffn.) sp. nov.

Bazzania curtiloba Schiffn. ms.

Pusillum, pallide-viride; caules parce ramosi; folia contigua vel laxe imbricata, oblique patula, ovata, 0.15—0.2 mm. longa, 0.1—0.13 mm. lata, bifida, lobis brevibus, truncatis, margine integro; cellulae in parte ventrali circa 17 μ latae, in parte dorsali circa 12 μ latae, parietibus incrassatis; foliola contigua vel subimbricata, trifida, lobis truncatis vel rotundatis; flores ignoti.

P h i l i p p i n e s: Mt. Batangan, Davao, Mindanao, 1888, O. WARBURG 14897 (V.), type of *Bazzania curtiloba* Schiffn. No other specimens are known.

The plants, which are pale green in color, are only 0.5—1 cm. long, with the dichotomies 1—3 mm. apart. Well-developed axes measure 0.1—0.12 mm. in width by 0.09—0.11 mm. in thickness. The cells of the dorsilateral rows of cortical cells (Fig. 22, *B*) are 25—30 μ in tangential width and 20—30 μ in radial width; those of the ventral rows are smaller, averaging about 15 μ in diameter. The latter are therefore of about the same size in cross-section as the medullary cells, which also average about 15 μ in diameter. The bounding walls of the cortical cells may attain a thickness of 6—8 μ; the other walls are thinner, but thickenings are present in the angles of the medullary cells and these may coalesce.

The leaves (Fig. 22, *A*) are contiguous to loosely imbricated and lie approximately in one plane, spreading at an angle of about 60 degrees. When seen from above the leaf-surface appears plane or slightly convex. Well-developed leaves are unsymmetrically ovate (Fig. 22, *C*) and measure 0.15—0.2 mm. in length by 0.1—0.13 in width. The dorsal margin curves outwardly, and the curvature at the base is scarcely greater than elsewhere; the ventral margin is straight or nearly so. In most of the leaves the sinus is narrow, and the divisions lie almost in contact; in some of the leaves, however, the divisions diverge up to an angle of perhaps 30 degrees. The bottom of the sinus varies from acute to rounded. The ventral division, which is truncate, consists of two rows of cells and, in most cases, is three or four cells long. Below the sinus the rows of cells are distinct to the base of the leaf, and each row is composed of four to six cells. The dorsal division, which is truncate or slightly indented, is three or four cells wide at the apex, more rarely two or five cells wide. At the level of the sinus this division, in most of the leaves, is four or five cells wide. The portion of the leaf below the sinus is eight or nine cells wide at the broadest part, narrowing to a width of perhaps six or seven cells at the base. The margin is entire throughout.

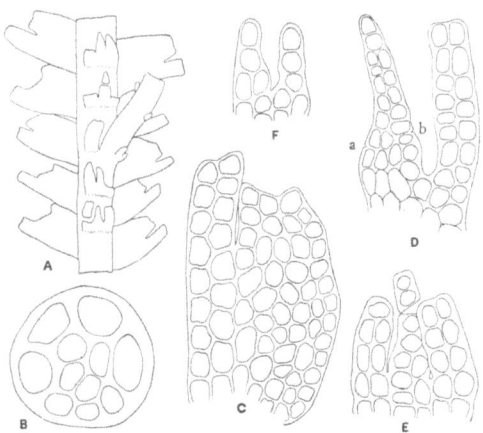

Fig. 22. *Acromastigum curtilobum* (Schiffn.) Evans. *A*. Part of plant, showing dorsiventral and flagelliform branches, ventral view, × 50. *B*. Crosssection of branch, × 225. *C*. Mature leaf, × 225. *D*. Young leaf, showing intact dorsal division, × 225. *E*. Underleaf, × 225. *F*. Leaf of flagelliform branch, × 225. The figures were drawn from the type-material.

The second row of cells from the ventral margin (Fig. 22, *C*), as in *A. bancanum*, might be interpreted as the vitta, although the cells of this row are scarcely distinguishable from the marginal cells or from the cells at the base of the third row from the margin. The cells of the vitta measure 20—25 µ in length by about 17 µ in width; in the dorsal part of the leaf the cells are smaller and more nearly isodiametric, averaging about 12 µ in diameter. The cell-walls are uniformly thickened, without evident trigones or pits, although the angles of the cavities are rounded. In the ventral part of the leaf the walls have a thickness of about 4 µ; in the dorsal part, of about 2 µ. The cuticle is densely but very faintly verruculose or striolate-verruculose.

The underleaves, which are contiguous to subimbricated and appressed to the axis, have a subquadrate outline, with straight or slightly bulging sides, and average about 0.08 mm. in length. The subparallel divisions (Fig. 22, *E*) are separated by narrow and acute sinuses and are, in most cases, two cells wide at the base and three to six cells long. Some of the divisions are two cells wide throughout and truncate at the apex; others are tipped with a single cell or with a row of two to four cells and rounded at the apex. On some of the underleaves the median division is longer than the others and may attain a length of 0.1 mm. The basal part, with occasional exceptions, is six cells wide and two cells high. The intermediate cells, to which the underleaves are attached, are scarcely larger than the basal cells. In most cases six intermediate cells are present, two of which lie at the base of each division; when only five intermediate cells are present, a single cell lies at the base of the median division. The cells and margins of the underleaves are essentially like those of the leaves.

The flagelliform branches are 0.06—0.08 mm. in diameter and bear scattered scale-like leaves. The latter (Fig. 22, *F*), which measure 0.08 mm. in length by 0.06 mm. in width when well developed, are subquadrate to oblong in outline. The acute sinus extends to below the middle, the erect divisions are two cells wide at the base and tipped with a row of two or three cells, and the part below the sinus is four cells wide and two cells high. Of course many of these scale-like leaves are even simpler in structure.

The leaves described on page 97 represent the condition at maturity, but such leaves are not intact. They have all lost the tip of the dorsal division and probably the tip of the ventral division as well. The loss takes place at a very early stage of leaf-development. Even in the leaves crowded together at the apex of a branch most of the divisions are already truncate. The apices of the leaves, as in other leafy hepatics, are the first parts to develop, and the ventral division apparently precedes the dorsal. As a result the cells of the divisions may be mature or nearly so, while the basal part of the leaf is still in a meristematic condition. This is illustrated by Fig. 22, *D*, which represents an immature leaf. It will be seen that the ventral division is truncate, but that the intact dorsal division is acute and tipped with a row of two cells. The separation of the caducous apical portion, which will take place along a line from *a* to *b*, has already begun. Owing to the scantiness of the material only a few branch-tips have been available for dissection, and all the ventral divisions observed were truncate at the apex. It is probable, however, that these divisions had lost their tips. This idea is supported by the frequent occurrence of shred-like fragments, which are found attached to the cell-walls at the truncate apex, and which might reasonably be interpreted as the vestiges of cells that had broken away. In the case of the dorsal division the line of cleavage is not definite in position. In some of the leaves, as shown in Fig. 22, *A*, it is situated at the bottom of the sinus; in other leaves, two or three cells above the bottom, as shown in Fig. 22, *C*.

The underleaves of *A. curtilobum* also show a tendency to fragment, but this tendency manifests itself later than in the leaves. It is not unusual, therefore, to find underleaves with one or more of their divisions intact. The lines of cleavage, which are sometimes indicated by slight indentations, break off the tips of divisions or entire divisions. Occasionally, in fact, a division may undergo two successive cleavages, the first cutting off the tip and the second the remainder. In extreme cases an underleaf may lose all of its divisions, so that nothing but the basal portion remains.

The caducous divisions of the leaves and underleaves in *A. curtilobum* are clearly organs of vegetative reproduction, although their germination has not yet been observed. The leaves and underleaves, in fact, may be interpreted as "Bruchblätter". In an organ of this type, according to the definition of CORRENS (7, p. 338), the line of cleavage is not indicated by definite histological features but may arise in various parts of the leaf. Bruchblätter are known in a good many species of the leafy hepatics, their occurrence has already been noted in *A. divaricatum*, and they are found also in several species of the allied genus *Bazzania*. In most cases, however, the entire leaf separates, except perhaps for a narrow and irregular basal band; and the condition found in *A. curtilobum*, in which only a division or the tip of a division separates, is much more unusual. A somewhat similar condition is found in *Plagiochila surinamensis* Sande Lac. of tropical America and in *Trichocolea fragillima* Herz. of the Philippines. In the *Plagiochila*, as shown by CARL (4, p. 149), the marginal teeth break off at or near the base, giving most of the mature leaves a ragged appearance; in the *Trichocolea*, as shown by HERZOG (17, p. 84. *f. 1, c, d*), the numerous capillary apices separate and leave truncate cells behind.

Bruchblätter of the type found in *A. curtilobum* are not infrequent in the mosses, and CORRENS lists a number of species which are characterized by the "fragility" of their leaves. The genus *Dicranum* is especially rich in species having this peculiarity, and CORRENS (7, p. 14) makes the interesting observation that the tendency of the leaf to break across decreases in passing from the apex toward the base. The cleavage takes place in the middle lamellae and does not cut through cells, except in the midrib. The line of cleavage, therefore pursues a somewhat zig-zag course.

Although the peculiar Bruchblätter of *A. curtilobum* will distinguish it at once from the other species of *Acromastigum*, it shows certain affinities with *A. divaricatum*. At the same time it could hardly be confused with this species, even in the absence of Bruchblätter. It is smaller in all its parts, the leaves and underleaves are composed of fewer cells, the leaf-cells are smaller, and the cortical cells of the branches are not only smaller but also have thicker walls.

18. **Acromastigum laevigatum** (Mitt.) sp. nov.

Herpocladium laevigatum Mitt. ms.

Pusillum, pallide-viride; caules parce ramosi; folia laxe imbricata, oblique bis subrecte patula, ovata, 0.3—0.35 mm. longa, 0.2—0.23 mm. lata, bifida, lobis triangulatis, acutis, margine integro; cellulae in parte ventrali 20—25 µ latae, in parte dorsali circa 18 µ latae, parietibus incrassatis; foliola dissita, trifida, lobis ligulatis vel subulatis, truncatis bis acutis; rami masculini ignoti; rami feminei breves; folia floralia paucijuga, intima ovata, bifida, lobis acuminatis, margine sparse denticulato vel breve ciliolato; perianthium fusiforme, circa 2.5 cm. longum, superne trigonum, ore constricto, laciniato; sporogonium ignotum.

B o r n e o: Sarawak, without date, A. H. EVERETT (N. Y.), type of *Herpocladium laevigatum* Mitt.; Mt. Matang, Sarawak, without date, A. H. EVERETT (N. Y.); K o e t e i, Boven Maratam Geb. Boekitmilli, 1898—99, AMDJAH, Exped. Nieuwenhuis 110 (Bog., Y.).

The plants are pale grayish green or yellowish green and the cell-walls are either colorless or very faintly pigmented. The stems are 1—2 cm. long, and the widely spreading dichotomies 2—5 mm. apart.

The branches, when well developed, are about 0.12 mm. in width and 0.1 mm. in thickness. The dorsi-lateral cortical cells (Fig. 23, *C*) measure 40—50 μ in tangential width and 20—30 μ in radial width.

FIG. 23. *Acromastigum laevigatum* (Mitt.) Evans. *A*. Part of plant, showing dorsiventral and flagelliform branches, ventral view, × 50. *B*. Part of branch ventral view, × 50. *C*. Cross-section of stem, × 225. *D, E*. Leaves, × 50. *F*. Cells from base of leaf *D*, ventral side, × 225. *G*. Dorsal base of same leaf, × 225. *H*. Apex of same leaf, × 225. *I*. Apex of dorsal lobe of leaf *E*, × 225. *J*. Apex of ventral lobe of same leaf, × 225. *K*. Underleaf, × 225. *L*. Same underleaf, × 225. The figures were drawn from the type-material.

In the section figured the ventral cortical cells are only 20—30 μ in tangential width and 10—20 μ in radial width, but in other sections the discrepancy in size between the dorsi-lateral cells and the ventral cells is much less. The medullary cells average about 20 μ in diameter, but two of the cells figured measure about 30 × 25 μ and thus exceed in size the ventral cortical cells. In most sections, however, the medullary cells are a little smaller than the ventral cells. The bounding walls of the cortical cells are, for the most part, 6—8 μ thick; the walls of the medullary cells are thinner but show local thickenings at the angles.

The leaves (Fig. 23, A, B) are loosely imbricated and spread at an angle of 60—80 degrees. They are not deflexed but lie approximately in a single plane. The upper surface is plane or nearly so in the dorsal part but more or less convex toward the ventral side. The ventral margin, in consequence, appears in profile view, either throughout its entire length or toward the base, when a branch is examined from below, The leaves are ovate and slightly unsymmetrical, measuring in well-developed examples 0.3—0.35 mm. in length by 0.2—0.23 mm in width. The dorsal base is rounded and arches to about the middle of the axis. Beyond the base the margin extends as a distinctly convex line to the apex of the dorsal division; the ventral margin is approximately straight. On explanate leaves (Fig. 23, D, E) the divisions spread at an angle of 45 degrees or somewhat less and are separated by an acute sinus about one third the length of the leaf. On attached leaves the ventral division, in many cases, curves gently forward. Both divisions are triangular and acute and are tipped with a single cell (Fig. 23, I, J) or with a row of two cells (Fig. 23, H). On some of the leaves the ventral division is slightly shorter than the dorsal, but in the majority of cases the divisions are subequal in length. The ventral division, however, is the narrower of the two; in a series of leaves examined the ventral divisions were two to five cells wide and the dorsal divisions four to six cells wide. The margin is entire throughout, except for the occasional presence of slight and irregular sinuations.

The cells near the ventral base of a leaf are larger than elsewhere but do not, in all cases, show an arrangement in definite longitudinal rows (Fig. 23, F). If the group of large cells is interpreted as the vitta, even when irregularly arranged, this structure is two or three cells wide at the base and is separated from the ventral margin by one or two rows of cells and from the dorsal margin by four to six rows. The cells of the vitta are 30—45 μ in length by 20—25 μ in width, the

cells along the dorsal margin average about 12 μ in width (Fig. 23, G),
and those in the divisions about 18 μ in diameter. The cell-walls are
uniformly thickened, and in the region of the vitta a thickness of
4 μ is sometimes attained. The cell-cavities, however, have rounded
angles; and, in some cases, vague indications of trigones, with the pits
obliterated or obscured by deposits of cell-wall substance, can be
demonstrated.

The underleaves are distant (Fig. 23, A, B) and closely appressed
to the axis. They are quadrate-orbicular in outline (Fig. 23, K), with
straight or slightly bulging sides, and measure (when well developed)
0.07—0.08 mm. in length by 0.07—0.1 mm. in width. The divisions
are separated by acute sinuses about two-thirds as long as the underleaves; they lie subparallel, but show a tendency to converge, especially while still attached. The divisions (Fig. 23, L), which are ligulate or
subulate in form, are two (or rarely three) cells wide at the base and
two and one-half to four and one-half cells long; they are tipped by
two cells side by side, by a single cell, or by two cells in a row. As a
rule the divisions are unequal, the median division being the smallest
and one of the lateral divisions being smaller than the other; but
there are many exceptions to this rule. The basal portion of an underleaf is, in most cases, six cells wide and one and one-half cells high.
The six intermediate cells, which often contain a dark substance. are
scarcely larger than the cells of the underleaves.

The flagelliform branches have a diameter of about 0.08 mm.; and
their scattered leaves, when well-developed, are about 0.08 mm. long
and 0.07 mm. wide. These leaves are divided to about the middle by
an acute sinus into two subulate divisions, tipped with a single cell
or a row of two cells. The divisions, in most cases, are two cells wide
at the base, and the basal portion of the leaf four cells wide and one
or two cells high.

The material of *A. laevigatum* is apparently destitute of male
branches but includes a few female branches, upon which the
perichaetial leaves are arranged in three or four series. Those of the
innermost series (Fig. 24, B) are ovate in general outline and measure
1.3—1.5 mm. in length by 0.6—0.7 mm. in width. They are bifid to
the middle or a little beyond, with long-acuminate, suberect or
connivent divisions, separated by an acute sinus. Each division is
tipped with a row of two or three narrow cells, and the margins of the

leaves are irregularly denticulate to short-ciliolate. The leaves of the
next outer series (Fig. 24, C) measure 0.9—1 mm. in length by 0.5—
0.6 mm. in width, and those of the third series (Fig. 24, D) 0.75—0.8
mm. by 0.4—0.5 mm. These leaves are relatively broader than those
of the innermost series and, in most cases, have fewer teeth. The leaf
of the second series shown in Fig. 24, C, is therefore somewhat a-
nomalous, since its teeth are larger than usual; one of them, in fact is

Fig. 24. *Acromastigum laevigatum* (Mitt.) Evans. *A*. Female branch with
perianth, × 27. *B*. Perichaetial leaf of innermost series, × 40. *C*. Perichaetial leaf of second series, × 40. *D*. Perichaetial leaf of third series, × 40.
E. Cells from side of perichaetial leaf *B*, × 225. *F*. Laciniae at mouth of
perianth, × 225. *G*. Apex of lacinia at mouth of perianth, × 225. The
figures were drawn from the type-material.

lobe-like in character. The cells of the perichaetial leaves, in most
cases (Fig. 24, *E*), are longer than broad and have uniformly thickened walls. The cuticle is striolate-verruculose.

The only perianth observed (Fig. 24, *A*) measured 2.5 mm. in
length and 0.7 mm. in diameter. It showed a narrowly ovate outline
and was deeply trigonous in the upper part, with a somewhat contracted mouth. The latter was surrounded by about ten irregular
and unequal laciniae (Fig. 24, *F*), each tapering to a slender cilium

composed of a single cell or of a row of two to five cells (Fig. 24, *G*). The sides of the laciniae are, in most cases, crenulate from projecting cells, but some of the marginal irregularities may be in the form of teeth or cilia. The cells are longer than broad and more delicate than those of the perichaetial leaves, at least in the upper part of the perianth. The cuticle is striolate, rather than striolate-verruculose.

The present species is apparently more closely related to *A. divaricatum* than to any of the other species. The plants in both are of about the same size, the color is very similar, the leaves are bifid and the underleaves trifid in about the same degree, and the vitta is about as indistinct in the one as in the other. The texture of *A. laevigatum*, however, is firmer than that of *A. divaricatum*, the tissues are less transparent, and the leaves are relatively broader, so that their form is ovate rather than ovate-rectangular. There are also other differences in the leaves that will help in distinguishing the species. In *A. divaricatum* the ventral margin appears in surface view, when a branch is examined from below; the ventral division is, as a rule, longer than the dorsal; and the leaf-cells in most cases show distinct trigones separated by thin areas. In *A. laevigatum*, on the other hand, the ventral margin throughout more or less of its extent appears in profile view, when a branch is examined from below; the ventral division in most cases is shorter than the dorsal or equals it in length; and the leaf-cells rarely if ever show distinct trigones.

19. **Acromastigum Cunninghamii** (Steph.) comb. nov.

Bazzania Cunninghamii Steph. Hedwigia 32: 205. 1893.
Mastigobryum Cunninghamii Steph. Spec. Hepat. 3: 540. 1909.

According to STEPHANI's original description the present species was based on No. 147 in the Kew Herbarium, a specimen collected by CUNNINGHAM at "Hay" Harbor, Straits of Magellan; according to the revised description in the Species Hepaticarum it was based on a specimen collected by CUNNINGHAM at Halt Bay, Straits of Magellan. Both of these specimens are represented in the MITTEN Herbarium, but the first is labeled "Gray" Harbor instead of Hay Harbor. This specimen, which consists of a single stem, agrees with the more abundant material from Halt Bay, upon which the figures in STEPHANI's Icones were based. The two specimens may be recorded as follows:

C h i l e: Gray Harbor, Straits of Magellan, November, 1868, A. CUNNINGHAM 147 (N. Y.), type of *Bazzania Cunninghamii* Steph.; Halt Bay, Straits of Magellan, April, 1868, A. CUNNINGHAM 96 (N. Y.). STEPHANI lists the species also from Hale Cove, Chile (40, p. 60), on the basis of specimens collected by SKOTTSBERG.

The plants are pale yellowish green, and the cell-walls in some cases show a very faint yellowish pigmentation, especially in the axial organs and along the ventral side of the leaves. The stems are 1—2 cm. long (up to 3 cm. long, according to STEPHANI), and the dichotomies 2—6 mm. apart. The ordinary vegetative branches, which show almost no dorsiventral flattening, are 0.12—0.18 mm. in diameter. The cortical cells, as seen in cross-section (Fig. 25, *B*), are 30—50 μ in tangential width by 30—45 μ in radial width, and the medullary cells measure about 18 μ in diameter. The bounding walls of the cortical cells may attain a thickness of 8—12 μ, but the walls of the medullary cells are much thinner, although they show well-developed and often coalescent thickenings at the angles.

The leaves are loosely imbricated (Fig. 25, *A*) and spread at an angle of 45—60 degrees. The dorsal part is plane or nearly so, but the ventral part is more or less convex. In many cases the convexity is sufficient to make the ventral margin appear in profile view, either throughout its entire extent or toward the base, when a branch is examined from below; in some cases, however, the ventral side appears in surface view. At the dorsal base the leaves are rounded to subcordate and may arch slightly beyond the middle of the axis. They are unsymmetrically ovate in form (Fig. 25, *C*) and measure, in the majority of cases, 0.4—0.5 mm. in length by 0.15—0.25 mm. in width. Beyond the base the dorsal margin extends as a straight or slightly incurved line to the apex of the dorsal division; the ventral margin, too, is approximately straight and lies subparallel with the dorsal margin in its outer half. The divisions are separated by an acute to rounded sinus about one-third the length of the leaf. On explanate leaves the divisions diverge at an angle of 30—45 degrees, but on attached leaves the ventral divisions, in many cases, curve forward. Both divisions are acute and tipped with a single cell or with a row of two cells. The ventral division (Fig. 25, *G*), which is linear, is not only narrower but slightly longer than the dorsal division (Fig. 25, *F*), which is triangular in form. In a series of leaves

FIG. 25. *Acromastigum Cunninghamii* (Steph.) Evans. *A*. Part of plant, ventral view, × 50. *B*. Cross-section of branch, × 225. *C*. Leaf, × 50. *D*. Cells from base of same leaf, ventral side, × 225. *E*. Dorsal base of same leaf, × 225. *F*. Dorsal division of another leaf, × 225. *G*. Ventral division of leaf from which *F* was drawn, × 225. *H, I*. Underleaves, × 50. *J*. Another underleaf, × 225. *K*. Leaf of flagelliform branch. The figures were drawn from specimens collected by CUNNINGHAM at Halt Bay, Chile.

examined the dorsal divisions were four to six cells wide at the base, whereas the ventral divisions were only two cells wide, not only at the base but throughout the greater part of their length. In rare instances, however, the ventral division is three cells wide above the base, and a leaf showing this peculiarity has been figured by STEPHANI in his Icones. The margin is entire throughout.

In most of the leaves three (or four) longitudinal rows of cells are present along the ventral side (Fig. 25, *D*), and the two outer rows extend into the ventral division. Here, as in *A. curtilobum*, the sub-marginal row may be interpreted as the vitta, although the cells of this row are scarcely distinguishable from the cells on either side. The cells of the vitta are 20—40 μ long and about 25 μ wide, and many of them are subquadrate in form. The vitta is separated from the dorsal margin (Fig. 25, *E*) by six to eight rows of cells; the marginal cells toward the base average about 14 μ in diameter and those of the submarginal row about 18 μ. The other cells in the dorsal part of the leaf are scarcely different in size from the cells in the ventral part, but their arrangement is more irregular. The cell-walls are everywhere firm and attain in many cases a thickness of 6 μ. The thickening appears uniform when a leaf is examined in surface-view, but pits may be visible when cells are examined in profile view. The cell-cavities are, for the most part, polygonal with rounded angles, and no trigones are apparent. In some cases, how-ever, especially along the margin, vague indications of trigones can be demonstrated. The cuticle is smooth throughout.

The underleaves (Fig. 25, *A*), which are contiguous to loosely imbricated, are appressed to the axis and therefore appear convex, when see from below. Their general outline is orbicular-quadrate (Fig. 25, *H*, *I*) with straight or slightly bulging sides, and well-developed examples are 0.2—0.25 mm. in length. The divisions, which are subparallel or slightly convergent, are ligulate to ovate in form, and vary from truncate or slightly retuse to acute at the apex (Fig. 25, *J*). They are two to four cells wide at the base and one to four cells wide at the apex, and an acute division may even be tipped with a row of two cells. On many of the underleaves the lateral divisions are wider than the median, but the three divisions may be subequal. The sinuses are one-half to two-thirds as long as the underleaf. They may be narrow and acute or broader and obtuse to rounded; and the divisions in consequence may be in contact or more or less separated.

The flagelliform branches average about 0.12 mm. in diameter and bear scattered, ovate, scale-like leaves. These, when well developed (Fig. 25, K), measure 0.2—0.25 mm. in length and are bifid about one-half with acute or truncate divisions. In some cases a more or less distinct basal dilatation is present on one or both sides. The sexual branches are still unknown.

STEPHANI compares this interesting species with A. laetevirens and states that it is distinguished by its plane and much narrower leaves, with narrower acuminate divisions. The narrow ventral divisions will serve also to distinguish A. Cunninghamii from almost all the other species with bifid leaves, which have so far been discussed. Among these it perhaps finds its closest allies in A. divaricatum and A. laevigatum. In A. divaricatum the leaves are as narrow as those of A. Cunninghamii or even narrower, and the ventral division may be only two cells wide throughout the greater part of its extent, although this is only exceptionally the case. The Indo-Malayan species, however, is the more delicate of the two, the ventral lobes in many cases are three or four cells wide at the base, the underleaves are smaller, the walls of the leaf-cells are thinner and more transparent, and the trigones are normally much more distinct. In A. laevigatum the plants are smaller than in A. Cunninghamii, the leaves are relatively broader, the leaf-divisions are broader and more abruptly pointed, and the underleaves are smaller and more simply constructed.

20. **Acromastigum obliquatum** (Mitt.) comb. nov.

Bazzania obliquata Mitt. in Stephani, Hedwigia 32: 211. 1893.
Mastigobryum obliquatum Mitt. in Stephani, Spec. Hepat. 3: 535. 1909.

In STEPHANI's original description of this fragile species, he gives the habitat as "Pacific Isles sine loco natali", and states that it was based on specimens received from G. DAVIES; but in his revised description of 1909 he gives the habitat as „Insula Salomonis (Penguin Expedition)." According to the data given in his unpublished Icones his figures were drawn from "Pacific" specimens collected by Davies, but it is stated that the species grows also on the Solomon Islands. The writer's description and figures were drawn from specimens in the Mitten Herbarium, which presumably represent a part of the type-material. These specimens may be recorded as follows:

Solomon Islands: without more definite locality, 1888, Captain Norman (N. Y.), probable type of *Bazzania obliquata* Mitt. No other stations for the species are definitely known to the writer. Herzog, however, lists it as a Polynesian endemic, which is not restricted to definite islands (16, p. 357).

The pale green plants of *A. obliquatum* are 1—2 cm. long, with the dichotomies 3—8 mm. apart; and the ordinary dorsiventral branches have a width of 0.12—0.15 mm. and a thickness of 0.01—0.12 mm. The cross-section of a well-developed axis (Fig. 26, *B*) shows that the cortical cells are 25—50 μ in tangential width by 20—30 μ in radial width and that the medullary cells average about 22 μ in diameter. The thickest walls are those bounding the cortical cells on the outside, and even here the thickness rarely exceeds 4 μ. The walls of the medullary cells, except for slight thickenings at the angles, remain thin.

The leaves are approximate to loosely imbricated and spread at an angle of about 90 degrees (Fig. 26, *A*). The upper surface is plane or slightly convex. In many of the leaves, however, the convexity is more pronounced along the ventral side in the outer part, so that the ventral division may appear in profile view, when a branch is examined from below. The shape of the leaf, therefore, can not always be seen clearly in attached leaves. When leaves are dissected off and spread out flat (Fig. 26, *C*, *D*) they are seen to be slightly falcate and to have a subrectangular or oblong-rectangular outline. Well-developed leaves measure 0.4—0.5 mm. in length by 0.2—0.25 mm. in width. At the dorsal base the margin is convexly curved but scarcely rounded, and the leaf barely reaches the middle of the axis. Beyond the base the margin extends as a straight or slightly convex line to the apex of the dorsal division. The ventral margin is subparallel with the dorsal and is straight or slightly concave throughout the greater part of its extent, usually becoming slightly convex in the outer part. The acute to obtuse sinus is one-fourth to one-third as long as the leaf and separates the divisions by an angle of 30—45 degrees. The divisions, in most cases, are obtuse and tipped with a single cell (Fig. 26, G); but the ventral division, in rare instances, is acute and tipped with a row of two cells. This division is not only

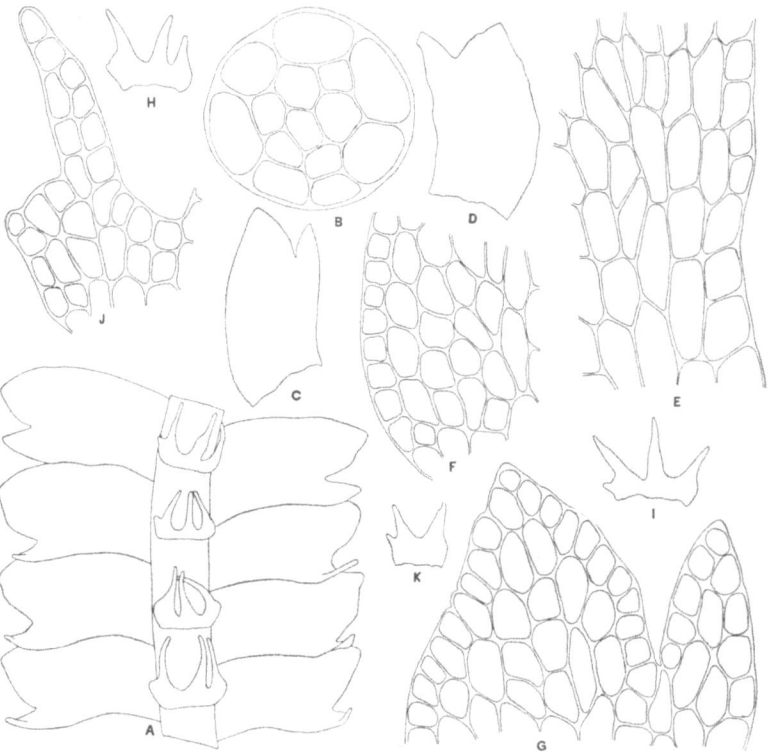

FIG. 26. *Acromastigum obliquatum* (Mitt.) Evans. *A*. Part of plant, ventral view, × 50. *B*. Cross-section of branch, × 225. *C, D*. Leaves, × 50. *E*. Cells from base of leaf *C*, ventral side, × 225. *F*. Dorsal base of same leaf, × 225. *G*. Apex of same leaf, the ventral division at right, × 225. *H, I*. Underleaves, × 50. *J*. Lateral division of underleaf *H*, × 225. *K*. Leaf of flagelliform branch, × 50. The figures were drawn from specimens collected by NORMAN in the Solomon Islands.

narrower than the dorsal but is, with rare exceptions, definitely shorter as well. In a series of leaves examined the dorsal divisions were eight to ten cells wide at the base, but the ventral divisions were only four or five cells wide. The margin is entire throughout, except for the fact that an occasional cell may project as a low and indistinct crenulation.

The indistinct vitta (Fig. 26, *E*), which in most leaves is two or three cells wide at the base, is separated from the ventral margin by one or two rows of cells and from the dorsal by seven to ten rows. The cells of the vitta are 40—60 μ in length by 20—30 μ in width; toward the dorsal margin (Fig. 26, *F*) the more nearly isodiametric cells average about 20 μ in diameter, and in the interior of the divisions the cells measure about 30 × 20 μ. Along the dorsal margin the cells average only 14 μ in width, and there is therefore a pronounced difference in size between the interior and marginal cells. The cell-walls are everywhere thin, but the walls in the vicinity of the vitta may be a trifle thicker than elsewhere. At the angles of the cells minute trigones with concave sides can be demonstrated, and pairs of trigones not infrequently coalesce. The cuticle is densely but minutely verruculose.

The underleaves (Fig. 26, *A*), which are distant to contiguous, spread slightly at the base and then curve abruptly forward, so that they lie parallel with the axis, without being appressed to it. They broaden out somewhat from the base, and their general outline might be described as broadly orbicular, particularly if attached underleaves are considered. Under these circumstances the divisions are subparallel or convergent, whereas the divisions of detached and explanate underleaves (Fig. 26, *H, I*) diverge more or less and thus modify the outline. Well-developed underleaves are 0.18—0.2 mm. in length by 0.2—0.3 mm. in width. They are deeply divided by obtuse to lunulate sinuses, two-thirds to three-fourths the length of the underleaf. The divisions are narrowly subulate, tapering from a base two cells wide (Fig. 26, *J*), and are tipped with a row of from two to seven cells (exclusive of the apical papilla). On most of the underleaves toward the base a lobe or tooth is present on one or both sides, and this may be blunt and rounded or sharp. In very rare instances the median division shows a similar but smaller tooth.

The flagelliform branches are 0.08--0.1 mm. in diameter and are therefore only a little more slender than the ordinary dorsiventral branches. Their leaves are broadly ovate (Fig. 26, *K*) and measure, when well developed, about 0.15 mm. in length. They are bifid to below the middle with acute to acuminate divisions and, in many cases, bear a lobe or tooth on one or both sides. These leaves thus resemble the underleaves, except that they are bifid instead of trifid. The sexual branches are unknown.

In *A. obliquatum* a species is met with in which the ventral division of a leaf is not only narrower than the dorsal but also definitely shorter. An approach to this condition has already been noted in *A. laevigatum*. In this species, too, the ventral division in many of the leaves is shorter than the dorsal. In the majority of the leaves, however, the divisions are subequal in length. Leaves in which the ventral division is shorter than the dorsal occur also in several of the other species that have been discussed, but do not in any case represent a constant feature.

The delicacy of the tissues in *A. obliquatum* is, in some cases at least, associated with a high degree of transparency, so that the longitudinal rows of cortical cells can be seen with unusual distinctness. The same transparency has been met with in *A. divaricatum*, but the two-species are not much alike in other respects. Aside from the differences in the leaf-divisions, with respect to their relative size, *A. obliquatum* is at once distinguished from *A. divaricatum* by differences in the underleaves and leaf-cells. In *A. obliquatum*, for example, the divisions of the underleaves are slender and long-pointed and the basal portion in many cases shows a lobe or tooth on one or both sides. The leaf-cells, moreover, are unusually thin-walled, and a rather marked contrast in size is apparent, particularly in the dorsal part of the leaf, between the interior cells and the marginal cells. This contrast is shown in Fig. 26, *F, G*. In *A. divaricatum*, on the other hand, the divisions of most of the underleaves are broad and truncate, and the basal portion is destitute of lateral outgrowths. The walls of the leaf-cells, moreover, are somewhat thicker than those of *A. obliquatum*, and no marked contrast in size is evident between the interior and marginal cells. This is shown by Fig. 20, *I*. The same features that separate *A. obliquatum* from *A. divaricatum* will distinguish it also from *A. laevigatum*. In this species the contrast in size between the marginal and interior cells is likewise slight (Fig. 23, *G*), and the cells are further distinguished by having firm and uniformly thickened walls.

21. **Acromastigum microstictum** (Mitt.) sp. nov.

Herpocladium microstictum Mitt. ms.

Pusillum, pallidum, flavo-viride; caules parce ramosi; folia dissita bis laxe imbricata, oblique bis subrecte patula, ovata, 0.3—0.4 mm. longa, 0.2—0.25 mm. lata, bifida, lobis inaequalibus, acutis, ventrali subulato, dorsali longiore et latiore, triangulato, margine integro; cellulae in parte ventrali 15—30 µ latae, in parte dorsali circa 19 µ latae, parietibus subincrassatis, trigonis nullis; foliola dissita, trifida, lobis subulatis, acutis vel acuminatis; flores ignoti.

Samoa Islands: mixed with a *Trichocolea*, without definite locality or date, T. POWELL (N. Y.), type of *Herpocladium microstictum* Mitt. Known only from the type-material.

The plants are pale yellowish green, with little or no pigmentation of the cell-walls, and the living portions are 1—2 cm. in length, with the dichotomies 2—5 mm. apart. Although the flagelliform branches conform to the *Acromastigum* type (Fig. 27, *A*) and although the dorsiventral leafy branches as a rule conform to the *Frullania* type and show an undivided leaf at the fork, an occasional leafy branch is intercalary in origin and springs from the axil of an underleaf. Well-developed branches are 0.12—0.15 mm. in width and 0.11—0.13 mm. in thickness. The cortical cells (Fig. 27, *B*) are 30—50 µ in tangential width by 20—30 µ in radial width, and the medullary cells average about 18 µ in diameter. The walls bounding the cortical cells on the outside are 4—8 µ thick, but the walls between the medullary cells are very thin, except for slight thickenings at the angles.

The leaves (Fig. 27, *A*) in most cases are distant but may be contiguous or even loosely imbricated. They spread at an angle of 60—90 degrees and lie approximately in a single plane. The upper surface is plane, except along the ventral side, where some convexity is apparent. On some of the leaves this convexity is great enough to make the ventral margin appear in profile view to a greater or less extent, when a branch is examined from below. The leaves show their unsymmetrically ovate form most clearly when dissected off and spread out flat (Fig. 27, *C*). Well-developed examples are 0.3—0.4 mm. long and 0.2—0.25 mm. wide, but considerably smaller leaves are not infrequently developed. The dorsal margin is uniformly convex in the basal half and then continues as a straight or nearly straight line to the apex of the dorsal division; the ventral margin is straight or slightly concave toward the base but, in many leaves, curves or bends forward in the outer part. On some of the

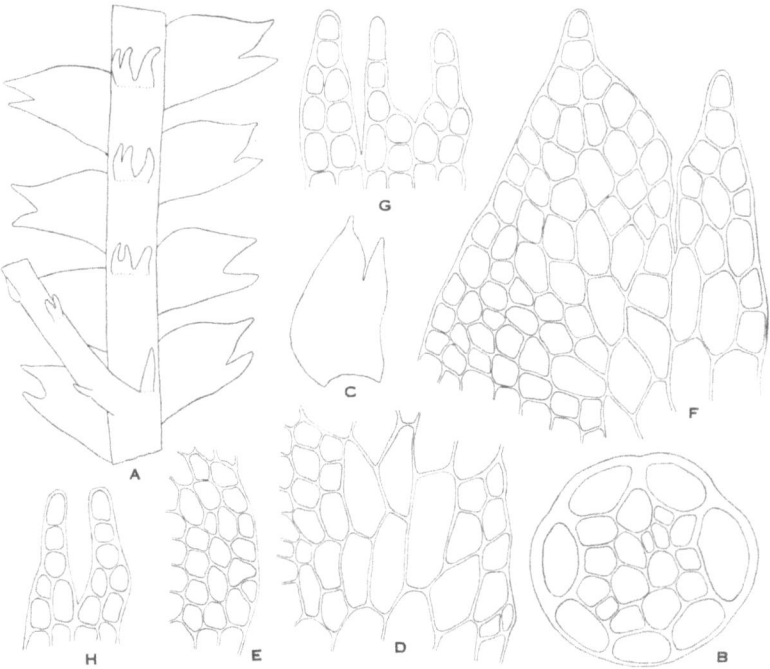

FIG. 27. *Acromastigum microstictum* (Mitt.) Evans. *A*. Part of plant, showing dorsiventral and flagelliform branches, ventral view, × 50. *B*. Cross-section of branch, × 225. *C*. Leaf, × 50. *D*. Cells from base of same leaf, ventral side, × 225. *E*. Dorsal base of same leaf, × 225. *F*. Apex of same leaf, × 225. *G*. Underleaf, × 225. *H*. Leaf of flagelliform branch, × 225. The figures were drawn from the type-material.

leaves the divisions are in contact; on other leaves they diverge at an angle, which may be as much as 30 degrees. The sinus is one-fourth to one-third as long as the leaf and is acute to obtuse at the bottom. Both divisions are acute and tipped with a single cell or with a row of two cells (Fig. 27, *F*). The ventral division, which is subulate, is both narrower and shorter than the triangular dorsal division. In a series of leaves examined the ventral divisions were only two or three cells wide at the base, but the dorsal divisions were five to seven cells wide. The margin is entire throughout, except for the presence of vague and irregular sinuations.

The vitta, in spite of its relatively large cells, is not clearly defined. In most leaves it is two or three cells wide at the base and is separated from the ventral margin by one or two rows of cells and from the dorsal margin by six to nine rows. The cells of the vitta are 30—70 μ in length by 15—30 μ in width and show considerable variation in form as well as in size (Fig. 27, *D*, *F*). Toward the dorsal margin (Fig. 27, *E*) the cells average about 19 μ in diameter, and the cells in the interior of the dorsal divisions measure about 28 × 17 μ. The walls throughout are slightly and uniformly thickened, although the angles of the cell-cavities are rounded. The cuticle is faintly but densely verruculose, becoming striolate-verruculose in the region of the vitta.

The underleaves are distant and appressed to the axis (Fig. 27, *A*), thus appearing slightly convex when seen from below. They are subquadrate in general outline (Fig. 27, *G*), with straight or slightly bulging sides, and measure, even when well developed, only 0.07—0.1 mm. in length. The divisions, which lie subparallel, are separated by acute to rounded sinuses, which extend to the middle or beyond. In most cases the divisions are subulate from a base two cells wide and are tipped with a single cell or with a row of two, three, or (rarely) four cells, not counting the apical papilla, which may be persistent. On most of the underleaves the two lateral divisions are larger than the median division and taper more gradually; the latter, in fact, may be only one cell wide throughout. The basal part of the underleaf is, in most cases, six cells wide and one and one-half or two cells high. The margins are entire or nearly so throughout.

The flagelliform branches are about 0.06 mm. in diameter and bear scattered, scale-like leaves, which attain a size of 0.1 × 0.06 mm., when well developed. These leaves are ovate or ovate-oblong in form (Fig. 27, *H*) and are deeply bifid by an acute to obtuse sinus. The divisions are subulate, two cells wide at the base, and tipped with a row of two or three cells. They are thus similar to the divisions of the underleaves. The sexual branches are unknown.

In *A. microstictum* another species is met with in which the ventral division of a leaf is not only narrower than the dorsal but also definitely shorter. It is, in fact, a close relative of *A. obliquatum*, which also shows this unusual feature. The two species have much in common and resemble each other in size, in color, in delicacy, and in many of the characters drawn from the cells of the axial organs and leaves. There are, however, important differences between them. In *A. microstictum*, for example, the apices of the leaf-divisions are acute and, in many cases, tipped with a row of two cells; there is only a slight difference in size between the interior cells and the marginal cells along the dorsal side of the leaf, as shown by Fig. 27, *F*; and the underleaves, which rarely if ever exceed 0.1 mm. in length, do not bear lateral teeth or lobes. In *A. obliquatum*, on the other hand, the leaf-divisions, in most cases, are obtuse and tipped with a single cell; there is a marked difference in size between the interior cells and the marginal cells along the dorsal side of the leaf; and the underleaves, which may attain a length 0.2 mm., show, in many cases, a distinct tooth or lobe on one or both sides.

22. **Acromastigum linganum** (De Not.) comb. nov.

Mastigobryum linganum De Not. Mem. Accad. Sci. Torino II. 28: 301. *pl. 29.* 1874.
Bazzania lingana Trevis. Mem. Ist. Lomb. 13: 414. 1877.

The species was based on material collected by Beccari on Mt. Linga in Borneo. The original figures of DE NOTARIS were drawn from this material, and it has served also for the figures in STEPHANI's Icones and for a series of original drawings by Professor SCHIFFNER, which is now in the possession of the writer. One of DE NOTARIS' figures has been reproduced by VERDOORN (42, *f. 40a*). The type-material of the species, which is now preserved in the herbarium of the University of Florence, agrees with the other specimens listed below. The species is still known only from Borneo.

B o r n e o: Mt. Linga, Sarawak, 1867, O. BECCARI 23 (F.), type of *Mastigobryum linganum* De Not.; Mt. Matang, Sarawak, without date, A. H. EVERETT (N. Y.); without definite locality, Sarawak, without date, native collector 1213 (Y.), distributed by the Bureau of Science, Manila, under the name *Mastigobryum Notarisii* Steph.;

Kapuas River, Dutch Borneo, 1923, collector not named (Herz., Y.), specimens sent by R. WEGNER to T. HERZOG.

The plants of *A. linganum* grow in depressed mats and are associated in some cases with other bryophytes. Their color is a dingy brownish green, paler in the younger parts and darker in the older. Although the cell-walls in the leaves remain unpigmented, those in the axial organs may acquire a yellowish brown coloration. In some of the specimens this coloration is less pronounced in the cortex than in the medulla, which then appears as a darker streak covered over by a paler translucent layer. The plants are 1.5—2 cm. long, so far as their living parts are concerned, and the dichotomies occur at intervals of 2—5 mm. Well-developed dorsiventral branches have a width of 0.18—0.2 mm. and a thickness of 0.16—0.18 mm. The cortical cells (Figs. 28, *B*; and 29, *A*) measure 40—60 μ in tangential width by 30—40 μ in radial width, and the medullary cells average about 14 μ in diameter. The bounding walls of the cortical cells are 6—10 μ in thickness; the walls of the medullary cells are much thinner but show, in many cases, triangular thickenings at the angles.

The leaves are approximate to loosely imbricated (Fig. 28, *A*) and spread at an angle of 90 degrees or a little less. Some of the leaves are more or less deflexed, but most of them lie approximately in a single plane. They are straight or slightly falcate and have an oblong outline, being of about the same width throughout the greater part of their length (Figs. 28, *C*; and 29, *B*). Well-developed leaves are 0.7—0.9 in length by 0.3—0.5 mm. in width. The dorsal margin is convexly curved at the base and then extends as an almost straight line to the apex of the dorsal division. The ventral margin also is almost straight to the base of the ventral division and then curves or bends forward, in some cases forming a more or less definite angle. The sinus, which is about one-third the length of the leaf, is acute to lunulate at the bottom and separates the divisions by an angle varying from 30 to 90 degrees. The divisions are acute and, in most cases, tipped with a single cell (Figs. 28, *F*; and 29, *D*). The dorsal division is always shorter and narrower than the ventral, and in most of the leaves the difference in size is very marked. In a series of leaves examined the dorsal divisions were three to eight cells wide at the base, whereas the ventral divisions were five to eleven cells wide. On some of the leaves the margin is entire or nearly so throughout, but on the majority a few scattered crenulations or denticulations are present. These may be nothing more than low projections of indi-

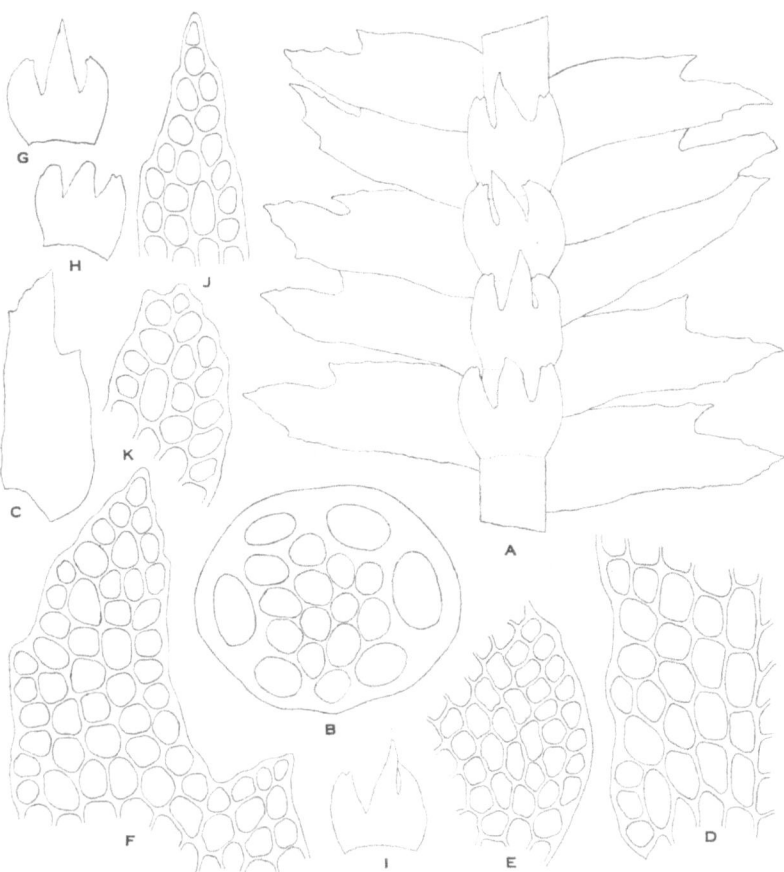

Fig. 28. *Acromastigum linganum* (De Not.) Evans. *A*. Part of plant, ventral view, × 50. *B*. Cross-section of branch, × 125. *C*. Leaf, × 50. *D*. Cells from base of same leaf, ventral side, × 225. *E*. Dorsal base of same leaf, × 225. *F*. Apical part of same leaf, × 225. *G-I*. Underleaves, × 50. *J*. Median division of underleaf shown in *G*, × 225. *K*. Lateral division of underleaf shown in *H*, × 225. The figures were drawn from specimens collected along the Kapuas River, Borneo.

vidual cells, as shown in Figs. 28, *F*, and 29, *D*, but may be slightly more complex. The teeth are perhaps more frequent along the lower side of the ventral division than elsewhere (Fig. 28, *A*).

The vitta is fairly distinct (Figs. 28, D; and 29, C) but fades out at about the middle of the leaf. On well-developed plants it is three to five cells wide at the base and is separated from the ventral margin by two or three rows of cells and from the dorsal margin by eight to fourteen rows. The cells of the vitta are 30—50 μ long and average about 15 μ in width; between the vitta and the dorsal margin (Fig. 28, E) the cells average about 17 μ in diameter and in the divisions about 20 μ. The cells show more or less distinct trigones with concave sides, and coalescences of pairs of trigones are not infrequent; otherwise the walls between the cells are thin or only slightly thickened. The superficial cell-walls, however, are distinctly thickened. This is shown clearly along the margins, where these walls appear in optical section. The cuticle is minutely and very faintly verruculose, but the verruculae are difficult to demonstrate, except perhaps at the margins.

The underleaves (Fig. 28, A) are approximate to loosely imbricated and are more or less appressed to the stem, so that they appear convex when seen from below. Their general outline is orbicular-quadrate (Figs. 28, G—I; and 29, E, F), and well-developed examples are 0.25—0.3 mm. in length. They are divided to about the middle by narrow sinuses, and the divisions lie subparallel. These divisions vary from ligulate to subulate and are four to seven cells wide at the base. The apex, in the majority of cases, is two to four cells wide and either truncate (Fig. 29, I) or emarginate (Figs. 28, K; and 29, G), with a hyaline papilla (or its vestige) in the depression and a more or less distinct rounded projection on each side. One of these projections, however, may be shorter than the other (Fig. 29, H) or obsolete, and under these circumstances the division becomes subacute or acute and may even be tipped with a row of two cells (Fig. 28, J). On some of the underleaves a blunt tooth (Fig. 29, J) is present on one or both sides, but on the majority the margin is entire or subentire throughout.

The flagelliform branches are considerably more slender than the ordinary leafy branches and measure, even when well developed, only 0.09—0.12 mm. in diameter. Their distant scale-like leaves may be as much as 0.15—0.18 mm. in length by 0.09—0.12 mm. in width but, in most examples, are much smaller. These leaves, which have an ovate outline, are bifid about one-third with acute to truncate divisions, separated by an acute to rounded sinus. The margin is entire or unidentate on one side.

A few male branches are present in BECCARI's material. These are more or less curved, and one example, which was 1.8 mm. in length,

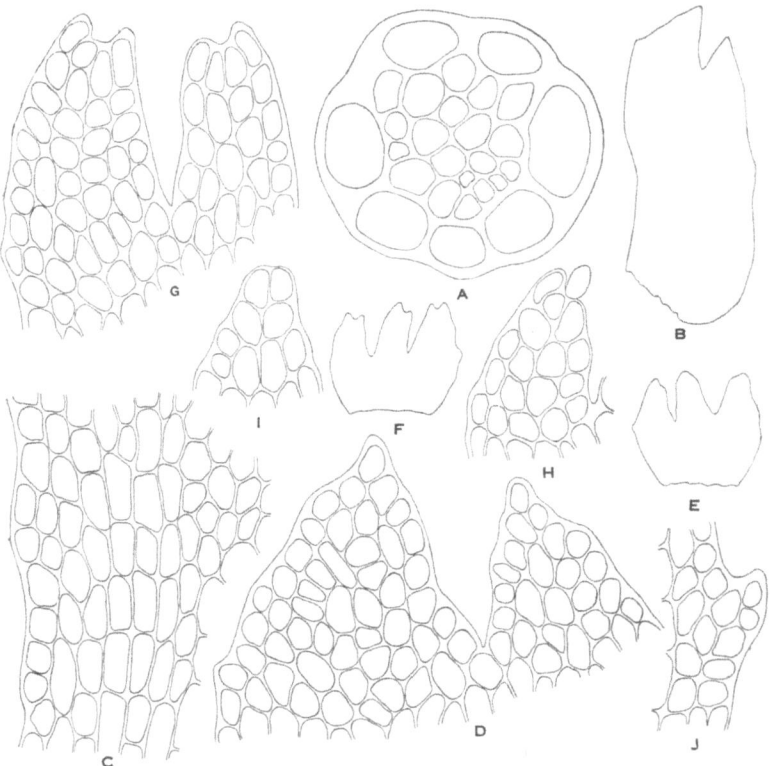

FIG. 29. *Acromastigum linganum* (De Not.) Evans. *A*. Cross-section of branch, × 225. *B*. Leaf, × 50. *C*. Cells from base of leaf, ventral side, × 225. *D*. Apical part of leaf shown in *B*, × 225. *E*, *F*. Underleaves, × 50. *G*. Median and lateral divisions of underleaf, × 225. *H*, *I*. Lateral divisions of underleaves, × 225. *J*. Lateral tooth of underleaf, × 225. The figures were drawn from specimens collected in Sarawak, Borneo, by a native collector, No. 1213

bore about twelve pairs of bracts. The latter (Fig. 30, *A*, *B*), although inflated, are somewhat compressed and are unequally complicate-bifid about one-third, with an arched and rounded keel. Explanate bracts (Fig. 30, *C*, *D*) are broadly ovate and measure about

0.4 mm. in length by 0.3 mm. in width. Both divisions are acute or short-acuminate and are tipped with a single cell (Fig. 30, E). The bracteoles are about 0.3 mm. in length by 0.25 mm. in width. They are ovate (Fig. 30, F) and bifid about one third with acute divisions and an acute to lunulate sinus. The margins of the bracts and bracteoles are entire or vaguely and sparingly crenulate from projecting cells. The cells toward the base are thin-walled, but those toward the apex have distinct trigones and, in some cases, thickened superficial walls. In Fig. 30, E, for example, the wall on the left-hand margin, which represents such a wall in optical section, is distinctly thickened, but the wall on the right-hand margin is thin. The cuticle is apparently smooth throughout.

The female branches bear three or four series of perichaetial leaves, which increase in size toward the archegonia. Those of the innermost series, which measure 1.5—2.2 mm. in length by 0.8—0.9 mm. in width, have an ovate form (Fig. 30, G, H) and are bifid to beyond the middle with a narrow sinus and acuminate divisions, tipped with two to four narrow cells in a row. The divisions are variously curved and contorted. In some cases they are deeply laciniate, and the leaves may thus acquire a multifid appearance (Fig. 30, H); in other cases they are difinitely bifid (Fig. 30, G). In addition to the laciniae the margins of the leaves bear scattered, irregular teeth, which vary from minute crenulations to short cilia. The leaves of the next outer series (Fig. 30, I) are shorter and less complex than those of the innermost series but are otherwise similar. The cells of which the perichaetial leaves are composed are more or less elongate and have slightly thickened walls and a faintly striolate-verruculose cuticle.

The perianths, which have a lanceolate or narrowly ovate outline, are terete below and deeply tricarinate above, with a laciniate and plicate mouth. One example in the BECCARI material had a length of 4 mm. and a diameter of 0.6 mm. According to DE NOTARIS the number of laciniae at the mouth of the perianth is definitely five, and five principal laciniae are shown in one of his figures (8, *pl. 29, f. 7*). Two of these laciniae, however, bear short secondary laciniae or cilia on their sides. According to the writer's observations five of the laciniae present are a little longer than the others and attain a length of 0.25—0.3 mm. Other laciniae, at the same time, are almost as long, so that it is somewhat misleading to state that only five are present. In all probability the number is subject to variation. The laciniae taper gradually from a base two to six cells wide to a capillary apex, composed of a row of two to four long cells. They vary not

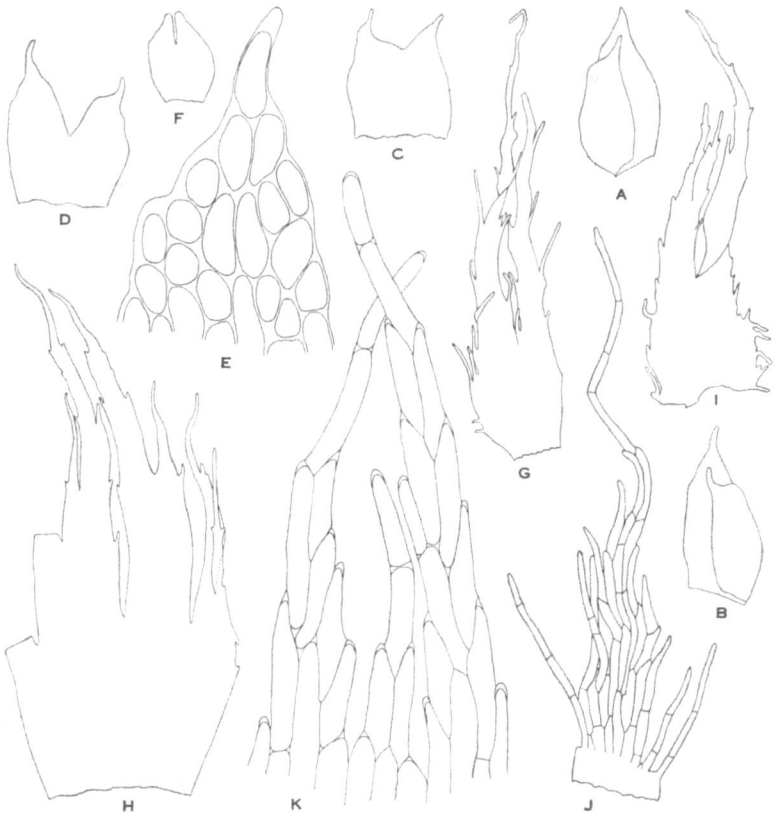

FIG. 30. *Acromastigum linganum* (De Not.) Evans. *A, B*. Male bracts, natural position, × 50. *C, D*. Male bracts, spread out flat, × 50. *E*. Apex of division of male bract, × 225. *F*. Male bracteole, × 50. *G*. Perichaetial leaf of innermost series, × 40. *H*. Perichaetial leaf from innermost series of another involucre, × 40. *I*. Perichaetial leaf from second series, × 40. *J*. Laciniae from mouth of perianth, × 100. *K*. Laciniae from mouth of another perianth, × 225. *A-F*. were drawn from the type-material; *G, I* and *J*, from the specimens collected along the Kapuas River, Borneo; and *H* and *K*, from No. 1213.

only in length and in the number of component cells but also in the charater of the margin, which may be minutely crenulate (Fig 30, *K*) or, at the other extreme, bear a series of irregular teeth or short cilia (Fig. 30, *J*). The cells of the laciniae are 60—90 μ long and about 12 μ wide, but the cells of the perianth below the laciniae are 70—100 μ long and 15—20 μ wide. The walls are thin but, in the laciniae at least, may show slight thickenings at their apical ends (Fig. 30, *K*). The cuticle is apparently smooth throughout.

Although *A. linganum* bears a certain resemblance to *A. Brotheri* and *A. echinatiforme* in general appearance, it can usually be distinguished without difficulty. In the majority of cases the dorsal leaf-division is distinctly shorter and smaller than the ventral division and, in extreme cases, looks more like a large tooth on the dorsal margin. In the other two species the difference in size between the divisions is much less pronounced. It is only in cases where the dorsal lobe of *A. linganum* is larger than usual (Fig. 29, *B*) that other differential characters must be utilized. Some of the best of these are derived from the leaf-cells and the leaf-margins. In *A. linganum*, for example, the cells have distinct trigones, separated by thin places in the walls; and the leaf margins, in many cases at least, develop distinct teeth. In *A. Brotheri* and *A. echinatiforme*, on the other hand, the cells have uniformly thickened walls, without distinct trigones or pits; and the leaf-margins are entire or nearly so throughout. Both species, moreover, are smaller and more deeply pigmented than *A. linganum*, and *A. echinatiforme* is further distinguished by its smaller cells.

23. **Acromastigum denticulatum** sp. nov.

Mediocre, flavo-fuscum, laxe caespitosum; caules 1—1.5 cm. longi, parce ramosi; folia laxe imbricata, oblique patula, ovato-oblonga, 0.5—0.6 mm. longa, 0.2—0.23 mm. lata, bifida, lobis anguste triangulatis, acutis, margine denticulato; cellulae in parte ventrali 15—20 μ latae, in parte dorsali circa 13 μ latae, parietibus incrassatis; foliola laxe imbricata, trifida, lobis ligulatis, apice retusis vel bidentatis; flores ignoti.

B o r n e o: Mt. Matang, Sarawak, without date, collector not named (N. Y.). Known only from the type-collection. The specimen, which is in the MITTEN Herbarium, is on a sheet bearing the specific name *elegantulum*. It does not agree, however, with *Mastigobryum elegantulum* De Not.

The plants are yellowish brown in dried condition and grew in depressed mats mixed with other bryophytes. The living portions were 1—1.5 cm. in length, and the successive dichotomies are 2—5 mm. apart. Well-developed leafy branches are 0.18—0.2 mm. in width and 0.15—0.17 mm. in thickness. The cortical cells, as seen in cross-section (Fig. 31, *B*), are 40—60 μ in tangential width by 35—50 μ in radial width, and the medullary cells have an average diameter of 23 μ. The outer walls of the cortical cells vary in thickness from 8 μ to 15 μ; the walls of the medullary cells are much thinner but develop thickenings at the angles.

The leaves (Fig. 31, *A*) are loosely imbricated and lie approximately in a single plane, although the upper surface may be slightly convex. They spread obliquely to widely, forming an angle of 60—80 degrees with the axis. Well-developed leaves are 0.5—0.6 mm. in length by 0.2—0.23 mm. in width. They are narrowly ovate-oblong in outline (Fig. 31, *C*), with a slight asymmetry, and are either straight or a little falcate. The dorsal base is rounded and arches to about the middle of the axis. Beyond the base the dorsal margin extends as an approximately straight line to the apex of the dorsal division. The ventral margin is subparallel with the dorsal for half or two-thirds its length and then curves gently forward. The sinus, which is one-third to one-half the length of the leaf, is sharply to bluntly pointed at the bottom. In some cases the narrowly triangular divisions are almost in contact, but they diverge on most of the leaves, and the angle of divergence may be as much as 30 degrees. Both divisions narrow gradually to sharp points and are tipped with a single cell or with a row of two cells. The ventral division, as a rule, is distinctly longer than the dorsal, but it may be just as narrow or even narrower. In a series of leaves examined the ventral divisions were four to six cells wide at the base and the dorsal six to nine cells wide. The margin (Fig. 31, *E—G*) is irregularly crenulate or denticulate from projecting cells. On some of the leaves the teeth are few and far between, but it is more usual to find them more abundant and closer together, especially along the dorsal side of the leaf and on the divisions.

The vitta (Fig. 31, *D*) is fairly well defined, and even the cells in the ventral division show a tendency to be arranged in longitudinal rows (Fig. 31, *G*). In the basal part of the leaf, where the vitta is three to five cells wide, it is separated from the ventral margin by two or

three rows of cells and from the dorsal margin by eight to eleven rows.
The cells of the vitta measure 20—40 μ in length by 15—20 μ in

FIG. 31. *Acromastigum denticulatum* Evans. *A*. Part of plant, ventral view,
× 50. *B*. Cross-section of branch, × 225. *C*. Leaf, × 50. *D*. Cells from base
of same leaf, ventral side, × 225. *E*. Cells from dorsal part of same leaf,
near base, × 225. *F*. Apex of dorsal division of same leaf, × 225. *G*. Apex
of ventral division of same leaf, × 225. *H-J*. Underleaves, × 50. *K*. Lateral
division of underleaf shown in *H*, × 225. *L*. Leaf of flagelliform branch,
× 50. The figures were drawn from the type-material.

width; the cells in the vicinity of the dorsal margin (Fig. 31, *E*)
average about 13 μ in diameter, and those in the divisions about
16 μ. The cell-walls throughout the leaf are uniformly thickened

without evident trigones; and in the vitta, where the thickening is most pronounced, the walls may attain a maximum of 4 μ in thickness. The walls bounding the free surfaces of the leaf are plane or nearly so, and the cuticle is apparently smooth throughout.

The underleaves, which are loosely imbricated (Fig. 31, *A*), are almost plane and lie lightly appressed to the axis. They are orbicular-quadrate in outline (Fig. 31, *H—J*), with slightly bulging sides, and well-developed examples measure 0.3—0.35 mm. in length. The narrow sinuses extend to below the middle and vary from acute to rounded. The divisions diverge slightly from one another in their natural position (Fig. 31, *A*), but the angles between them may increase under pressure (Fig. 31, *I*). Most of the divisions are five to seven cells wide at the base and four cells wide in the apical part. They thus taper slightly, although their general form is ligulate. The apex is nearly always deeply retuse or sharply bidentate (Fig. 31, *K*), and in the latter case each tooth consists of a single cell or of two cells in a row. The margins of the underleaves are much like those of the leaves; in rather rare instances, however, a lobe-like tooth is present on one or both sides (Fig. 31, *I, J*).

The flagelliform branches have a diameter of 0.09—0.12 mm. and bear contiguous to distant scale-like leaves. The latter may attain a size of 0.2 × 0.12 mm. and have an ovate form (Fig. 31, *L*). They are bifid to beyond the middle, and their tapering divisions may be acute at the apex or retuse. Their margins are sparingly and vaguely crenulate. The sexual branches of the species are still unknown.

It will be sufficient, for the present, to compare *A. denticulatum* with *A. linganum*. The species are similar in size and in color, the leaves are similar in general form, and the leaf-divisions agree in being acute. There are certain features, however, which are fore-shadowed (as it were) in *A. linganum*, but which are more fully realized in *A. denticulatum*. This is true, for example, of the denticulations of the leaves and of the tooth-like structures at the apices of the underleaf-divisions. In *A. linganum* the teeth are never numerous and some of the leaves are entire throughout, although such leaves, in well-developed plants, are always associated with denticulate leaves. The underleaf-divisions, similarly, if emarginate at the apex, show rounded projecting cells on each side. In *A. denticulatum*, on the contrary, every leaf has some teeth, and the most characteristic leaves have numerous teeth, even if some of the teeth are nothing more than low crenulations. The underleaf-divisions, similarly, which are nearly always emarginate, show distinct teeth at the apical angles, and these teeth may even be formed of a row of two cells. Aside from these differences *A. denticulatum* can be distinguished from *A. linganum* by its subequal leaf-divisions and by its smaller leaf-cells with uniformly thickened walls. In *A. linganum* the dorsal division, in nearly every leaf, is distinctly smaller than the ventral, and the leaf-cells have distinct trigones.

24. **Acromastigum inaequilaterum** (Lehm. & Lindenb.) comb. nov.

Jungermannia inaequilatera Lehm. & Lindenb. in Lehmann, Pug. Plant. 6: 56. 1834.
Mastigobryum inaequilaterum Lehm. & Lindenb. in G. L. & N. Syn. Hepat. 218. 1845.
Jungermannia dirhyncha Tayl. ms., in part, *l.c.*, as synonym of *M. inaequilaterum.*
Mastigobryum elegantulum De Not. Mem. Accad. Sci. Torino II. 28: 300. *pl. 28.* 1874.
Bazzania inaequilatera Trevis. Mem. Ist. Lomb. 13: 414. 1877.
Bazzania elegantula Trevis. *l.c.*
Mastigobryum Notarisii Steph. Hedwigia 25: 241. 1886.
Bazzania Notarisii Schiffn. Conspect. Hepat. Archipel. Indici 167. 1898.

The original material of *Jungermannia inaequilatera* was collected by WALLICH and came from two different localities, Nepal and Singapore. The authors of the species distinguished the Singapore plants as var. β. *minor*, and this variety was soon raised to specific rank by GOTTSCHE, under the name *Mastigobryum echinatum* (14, p. 218). Subsequent writers have recognized the validity of this species without question. It will be shown below, however, that the line between *A. inaequilaterum* and GOTTSCHE's species is not always easy to draw. The segregation of the var. *minor* as a distinct species left Nepal as the only definite station for *A. inaequilaterum*. TAYLOR's manuscript species, *J. dirhyncha*, to be sure, was based on specimens

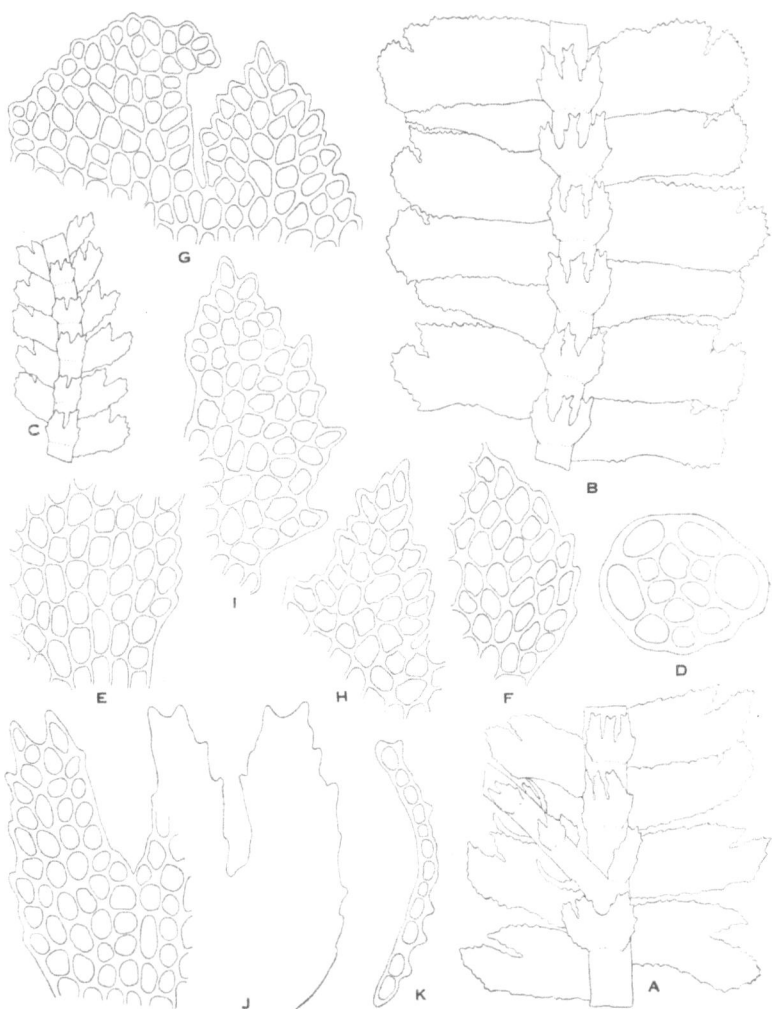

Fig. 32. *Acromastigum inaequilaterum* (Lehm. & Lindenb.) Evans. *A*. Part of plant, showing dorsiventral and flagelliform branches, ventral view, × 50. *B*. Part of plant, ventral view, × 50. *C*. Slender branch, ventral view, × 50. *D*. Cross-section of slender branch, × 225. *E*. Cells from base of leaf, ventral side, × 225. *F*. Dorsal base of leaf, × 225. *G*. Apical part of leaf, × 225. *H*. Dorsal division of leaf from which *E* was drawn, × 225. *I*. Ventral division of same leaf, × 225. *J*. Underleaf, × 225. *K*. Cross-section of underleaf, × 225. The figures were drawn from specimens collected by Bartlett near the Aek Kanopan, Sumatra, Nos. 7075 and 7077.

which were said to have come from Tasmania, but TAYLOR himself threw doubt on the correctness of this statement; and there is certainly no evidence that *A. inaequilaterum* has since been collected anywhere in the Australasian region. It was, nevertheless, soon reported from other parts of Indo-Malaya, and its known range now includes several of the East Indian islands. The following specimens have been examined by the writer:

A m b o i n a: without definite locality or date, J. E. TEYSMANN (H., N. Y.), cited by Stephani (39, p. 536) as *M. echinatum.*

B a n k a: near Baroe, Batoe-roesak, 1858, S. KURZ (L.); Mt. Maras, without date, VAN DIEST (L.). Both of these stations are listed by VAN DER SANDE LACOSTE (28, p. 301), in part under *M. echinatum.*

B o r n e o: without definite locality, Sarawak, 1867, O. BEC-CARI (F.), type of *Mastigobryum elegantulum* De Not.; Sibu, Rigang River, Sarawak, without date, A. H. EVERETT (N. Y.); Mt. Bangok, Sarawak, without date, A. H. EVERETT (N. Y.); without definite locality, Sarawak, without date, native collector 1579 (Y.), distributed by the Bureau of Science, Manila, under the name *M. echinatum*; without definite locality, Labuan, without date, J. MOTLEY (N. Y.), cited by MITTEN as from Borneo (22, p. 104), under the name *M. echinatum*; Mt. Kina Bala, North Borneo, without date, F. W. BURBIDGE (N. Y.); near Pontianak, Dutch Borneo, without date, VAN OORSCHOT (H., L.); Sampit, Mendawe River, Dutch Borneo, 1921, K. HENDRICK (Herz., Y.); Bukit Mehipit, Dutch Borneo, 1924, H. WINKLER 3350 (Herz., Y.), listed by HERZOG (18, p. 191) as *M. Notarisii*; Koeala-Koeroen, Dutch Borneo, 1924, H. LAMPMANN (Herz., Y.). Reported also from Borneo by STEPHANI (39, p. 536) on the basis of specimens collected by GREBE and by TEYSMANN.

G r e a t   K a r i m o n   I s l a n d: without definite locality or date, H. Fox (N. Y.).

J a v a: Mt. Salak, 1894, V. SCHIFFNER (V., Y.).

M a l a c c a: Mt. Ophir, Johore, without date or collector's name (N. Y.), specimen from the Griffith Herbarium, distributed from Kew; same locality, without date, H. N. RIDLEY 723 (N. Y.), cited by STEPHANI (39, p. 536) as from Malacca; same locality, 1898, collector not named (N. Y.); same locality, 1930, F. VERDOORN 79 (V., Y.); Koatah Tingih, Johore, 1930, F. VERDOORN 172, 179 (V., Y.);

Bukit Timah, Singapore, 1897, H. N. RIDLEY 372 (N. Y.); same locality, 1893, V. SCHIFFNER 320 (V., Y.); same locality, 1930, F. VERDOORN 2, 9, 10, 12 (V., Y.); Negri Sembilan, 1898, H. N. RIDLEY 736 (N. Y.); Pulu Penang, without date, H. N. RIDLEY (N. Y.); Perak, 1874, C. CURTIS (H.).

Natuna Islands: Great Natuna, without date, W. MICHOLITZ (H.), cited by STEPHANI (39, p. 536) as *M. echinatum*.

Nepal: without definite locality or date, N. WALLICH (B., N. Y., Y.), type of *Jungermannia inaequilatera* Lehm & Lindenb.; also without definite locality or date, T. SCOULER (N. Y.), cited by GOTTSCHE, LINDENBERG & NEES VON ESENBECK (14, p. 218).

New Guinea: Galewo Strait, 1875, F. NAUMANN (H.), cited by SCHIFFNER (29, p. 17); additional specimens collected by NAUMANN at McCluer Bay on the same island are likewise cited. SCHIFFNER, at a later date, threw doubt on the accuracy of these records (32, p. 161) and, in his herbarium, transferred the specimens from Galewo Strait to *Bazzania Notarisii*. This species, however, in the present revision, is included among the synonyms of *A. inaequilaterum*.

Nicobar Islands: Kamorta, 1890, S. KURZ (H.), cited by STEPHANI (39, p. 436) under *M. echinatum*.

Sumatra: Battang Lekoo, without date, J. E. TEYSMANN (L.), cited by VAN DER SANDE LACOSTE (28, p. 301); Engano Island, 1894, MODIGLIANI 145 (H.), cited by STEPHANI (39, p. 536); near the Aek Kanopan, Landoet Concession, Koealoe, 1927, H. H. BARTLETT 7075, 7077 in part (Mich., Y.); near Bilah Pertama, 1928, RAHMAT SI TOROES 240 (Mich., Y.). Reported also from Sumatra by STEPHANI (39, p. 536), on the basis of specimens collected by WEYERS.

The various specimens which are here referred to *A. inaequilaterum* exhibit considerable diversity. Some of them agree in all essential respects with WALLICH's original material from Nepal. Others diverge from this material in various ways and, in some cases, to such an extent that the propriety of including them under *A. inaequilaterum* is perhaps open to question. Among these divergent specimens the form which DE NOTARIS segregated out as *M. elegantulum* is represented and also an equally distinct form which has been repeatedly referred to *M. echinatum*. The writer feels convinced, however, that these extreme forms are connected with the typical

Nepal plants by intergrading conditions. The description which follows emphasizes the general features of the species but takes into account also the more important variations.

The plants grow in depressed mats, in many cases mixed with other bryophytes, and vary in color from a pale dingy green to a dull brownish. The living portions are mostly 1—1.5 cm. long, and

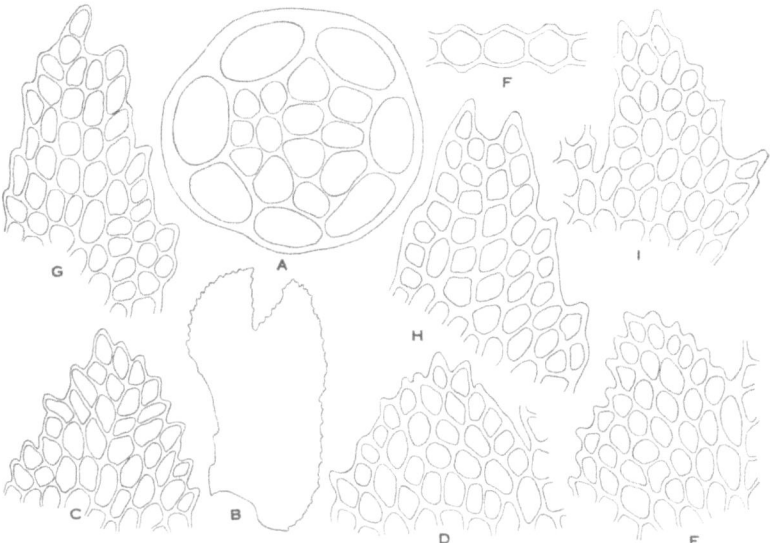

FIG. 33. *Acromastigum inaequilaterum* (Lehm. & Lindenb.) Evans. *A*. Cross-section of branch, × 225. *B*. Leaf, × 50. *C-E*. Dorsal divisions of leaves, × 225. *F*. Cross-section of leaf in outer part, × 300. *G-I*. Lateral divisions of underleaves, × 225. *A-C*, *F*, and *G* were drawn from the type-material of *Jungermannia inaequilatera*; *D* and *H*, from specimens collected by LAMP-MANN at Kocala-Koeroen, Borneo; and *E* and *I*, from specimens collected by SCHIFFNER on Mt. Salak, Java.

the widely spreading dichotomies occur at intervals of 2—3 mm. In rare instances an ordinary leafy branch arises in the axil of a normal underleaf, but the flagelliform branches conform throughout to the *Acromastigum* type and show an incomplete underleaf at the base. The ordinary leafy branches show a slight dorsiventral compression and measure, when well developed, 0.12—0.18 mm. in width by 0.09—0.15 mm. in thickness. The cortical cells (Figs. 32, *D*; 33, *A*; and 35, *A*) have a tangential width of 30—60 μ and a radial width of 25—40 µ, whereas the medullary cells average about 20 µ in diameter. As in other species the smaller cortical cells are those derived from the ventral segments. The walls of the cortical cells, which show little or no pigmentation, are strongly thickened, but the medullary cells are thin-walled except at the angles, where slight thickenings may be demonstrated. The thickness of the outer cortical walls in robust specimens is 10—14 µ but may be only 6—8 µ in delicate specimens.

The leaves are loosely to closely imbricated (32, *A*, *B*) and spread at an angle approximating 90 degrees. In the majority of cases they are plane or nearly so along the dorsal side but slightly convex in the ventral part. Plants are occasionally met with, however, in which some of the leaves at least are revolute in the outer part, and in extreme cases the revolute portion may involve both of the divisions and extend well toward the base along the ventral side. Well-developed leaves are 0.4—0.85 mm. long and 0.2—0.35 mm. wide, and have a subrectangular form (Figs. 33, *B*; 34, *A*; and 35, *B*), in some cases with a slight falcate curvature. The sides of the rectangle are formed by the dorsal and ventral margins, which are subparallel to beyond the middle and are either straight or a little curved; the outer end is formed by the outer part of the ventral margin, which curves or bends forward; and the inner end by the line of attachment and the margin of the rounded base. The narrow and acute sinus, which extends obliquely inward from the upper and outer angle of the rectangle, is one-seventh to one-foruth the length of the leaf and, in the more characteristic examples, slightly separates the two divisions. The latter are triangular and in most cases subacute, being tipped with a single cell. The dorsal division, which is straight and directed outward (Figs. 32, *G*, *H*.; 33, *C—E*; 34, *D*; and 35, *F*, *G*), is six to eleven cells long and five to ten cells wide at the base; the larger ventral division (Figs. 32, *G*, *I*; 34, *E*; and 35, *H*), which is directed forward, is six to sixteen cells long and six to eleven cells wide at the base.

Of course the account just given, which is somewhat rigid, applies to more or less typical leaves, but examples are often met with which differ in one way or another from the typical condition. The sinus, for example, may be subacute, obtuse, or even rounded and broad, rather than narrow. Under these circumstances the lobes will diverge more or less widely. The ventral margin, too, may be only gently curved in the outer part, causing the ventral division to be directed outward, rather than forward, and thus changing the contour of the entire leaf. The divisions, finally, may be obtuse or, in the case of the dorsal division, more sharply acute and tipped with a row of two cells. A considerable range of variation, in fact, is often found on an individual plant, especially if some of the branches are more robust than others.

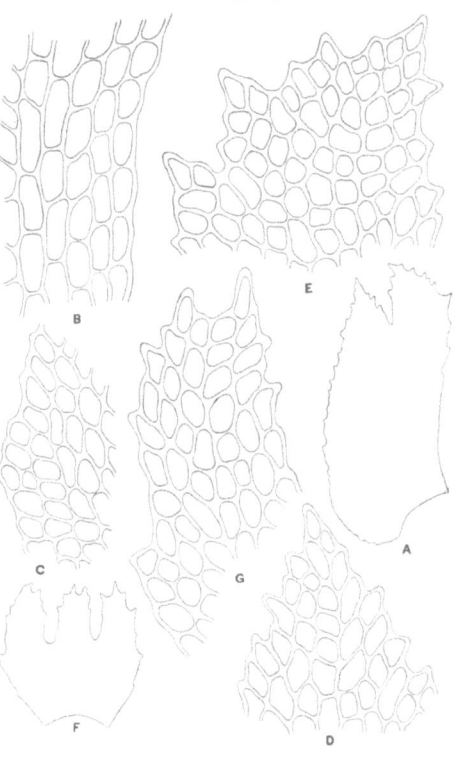

Except close to the base the leaf-margin is minutely crenulate or denticulate from projecting cells. The crenulations or denticulations vary considerably in height and are irregular in their distribution, being sometimes crowded and sometimes more or less scattered, owing to the fact that some of the marginal cells may fail to project. On some examples a few of the teeth are longer and usually sharper than the others (Figs. 32, *I*; and 34, *E*). These longer teeth are

FIG. 34. *Acromastigum inaequilaterum* (Lehm. & Lindenb.) Evans. *A*. Leaf, × 50. *B*. Cells from base of another leaf, ventral side, × 225. *C*. Dorsal base of same leaf, × 225. *D*. Dorsal division of same leaf, × 225. *E*. Ventral division of same leaf, × 225. *F*. Underleaf, × 50. *G*. Lateral division of another underleaf, × 225. The figures were drawn from specimens collected by HENDRICK at Sampit, Borneo; they are in essential agreement with the type-material of *Mastigobryum elegantulum* De Not.

mostly confined to the outer part of the ventral margin and may be unicellular or multicellular. The unicellular teeth are much like the ordinary marginal denticulations, except that they project beyond them. The multicellular teeth, although rarely consisting of more than three cells (Fig. 34, *E*), are a little more complex and often bear low crenulations on the sides.

The vitta (Figs. 32, *E*; 34, *B*; and 35, *C*, *D*) is poorly defined and consists of a band two or three cells wide in the basal part of the leaf. It is separated from the ventral margin by two to four rows of cells and from the dorsal margin by eight to fourteen rows. The cells of the vitta are mostly 20—30 μ long and about 15 μ wide but may, in certain forms, attain a length of 40 μ and a width of 20 μ. In the dorsal part of the leaf (Figs. 32, *F*; 34, *C*; and 35, *E*), between the vitta and the margin, the cells average about 15 μ in diameter; in the divisions they are a little larger and average about 18 μ in length by 16 μ in width. The cell-walls, for the most part, are uniformly thickened and show neither trigones nor pits. On some leaves, however, especially in the divisions, indistinct trigones can be demonstrated, but even in such cases the pits separating the trigones are more or less obliterated by depositions of cell-wall substance. The cavities of the cells are rounded at the angles.

On the dorsal surface of the leaf the cell-walls usually protrude in the form of low domes or cones. These protuberances, toward the base of the leaf, are restricted to the dorsal surface but are often prominent on both surfaces in the apical portion. The walls in some cases are uniformly thickened, giving the protuberances an even contour when seen in cross-section (Fig. 33, *F*). In the majority of cases, however, the wall is thicker in the middle. When the difference in thickness is slight, the apex may appear vaguely truncate (Fig. 33, *F*, at right); when the difference is more pronounced, a central tubercle, or verruca, is developed. These tubercles vary in height and have the form of a hemisphere or short rounded cylinder (Fig. 35, *F*, *I*), which is not necessarily sharply defined. In cells with unusually large tubercles the upper boundary of the cell-cavity becomes more or less flattened. The cuticle of the leaf is minutely and densely verruculose, but the verruculae may be difficult to demonstrate. They are most conspicuous, as a rule, along the margin and on the surface of the tubercles.

The underleaves (Fig. 32, *A*, *B*) are approximate to loosely imbricated and are more or less convex when seen from below. Their subquadrate form, somewhat narrowed toward the base, shows clearly in examples dissected off and spread out flat (Figs. 34, *F*; and 35, *J*). Well-developed underleaves are 0.2—0.35 mm. long. The narrow sinuses, which are acute to rounded, extend to the middle or a little beyond, and the subequal divisions are only slightly divergent. These divisions, in most cases, are four to nine cells wide at the base and two to four cells wide at the apex (Figs. 32, *J*; 33, *G—I*; 34, *G*; and 35, *K*). The latter is normally emarginate, with a sharp uni-cellular tooth on each side. In some cases, however, one of these teeth is smaller than the other or wholly obsolete, and the apex in conse-quence becomes acute. The margins of the underleaves are minutely crenulate or denticulate, much as in the leaves, and a larger tooth may be present on one or both sides (Figs. 33, *I*; and 35, *K*). The cells of the underleaves are essentially like the leaf-cells, but the convexities of the free walls are restricted to the ventral surface and the same thing is true of any cuticular tubercles that may be present (Fig. 32, *K*).

The flagelliform branches, although usually simple, may be sparingly branched. They are 0.09—0.12 mm. in diameter and bear approximate to distant, scale-like leaves, which may attain a length of 0.15 mm. These leaves, when well developed, are ovate, bifid to about the middle with acute divisions, and crenulate or denticulate along the margin (Fig. 36, *A*, *B*). Toward the tips of the branches, especially if rhizoids are present, the scale-like leaves become irregular and greatly reduced in size. In an extreme case a leaf of this character may consist of only four cells, two forming the base and the other two representing the divisions. The cells of the scale-like leaves are essentially like those of the underleaves, and their convex bounding walls are similarly restricted to the surface turned away from the branch. If a plant bears tubercles these may often be more easily demonstrated on the leaves of the flagelliform branches than elsewhere, because many of them show in optical section.

Although the distinctions between the vegetative and flagelliform branches are usually well marked, this is not always the case. A vege-tative axis, for exemple, may assume a flagelliform character, either abruptly or gradually; and a change in the opposite direction, although less frequent, may occasionally be observed. Poorly de-veloped vegetative branches, moreover, sometimes show interesting reductions. In the branch figured (Fig. 32, *C*) the leaves are much

smaller than ordinary robust leaves, and the underleaves are bifid
instead of trifid. Such a branch, to a certain extent, represents a
transition between vegetative and flagelliform branches.

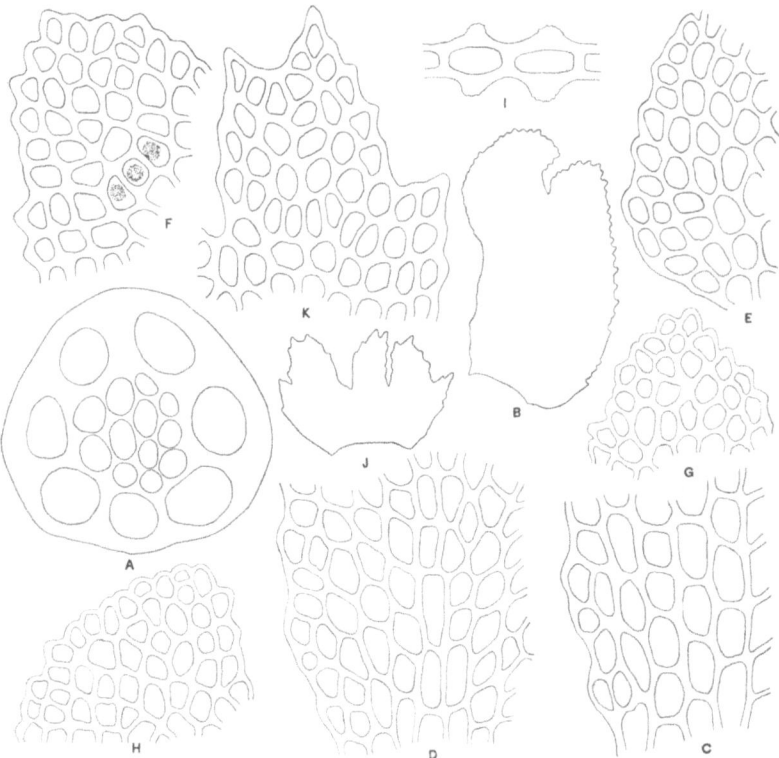

FIG. 35. *Acromastigum inaequilaterum* (Lehm. & Lindenb.) Evans. *A*. Cross-
section of branch, × 225. *B*. Leaf, × 50. *C, D*. Cells from bases of leaves,
ventral side, × 225. *E*. Dorsal base of leaf, × 225. *F, G*. Dorsal divisions
of leaves, × 225. *H*. Ventral division of leaf, × 225. *I*. Cross-section of leaf
in outer part, × 300. *J*. Underleaf, × 50. *K*. Lateral division of underleaf,
× 225. *A-C, E, F, I, K ,& K* were drawn from specimens collected by VER-
DOORN on Mt. Ophir, Malacca, No. 79; *D, G*, and *H*, from specimens collected
by RIDLEY in the same locality, No. 723.

The male branches normally arise in the axils of underleaves on
ordinary vegetative branches but occasionally occur on flagelliform
branches. They are more or less curved, sometimes so strongly so
that the apex touches the base. A robust branch, such as the one

figured (Fig. 36, C) is about 0.3 mm. in diameter and may attain a length of 1 mm. The closely imbricated monandrous bracts are, on some of the branches, in twelve to fifteen pairs but may be less numerous. They are inflated and complicate-bifid (Fig. 36, D), with a rounded and arched keel and acute to acuminate divisions. The explanate bract figured (Fig. 36, E) is 0.4 mm. long and 0.25 mm. wide and shows an ovate outline. The bracteoles, which are convex on the lower surface, are smaller than the bracts but agree with them in being ovate and bifid. A bracteole of average size (Fig. 36, G) is about 0.2 mm. long and 0. 1 mm. wide. The margins of the bracts and bracteoles are minutely and irregularly denticulate from projecting cells (Fig. 36, F), the walls are slightly thickened and may show vague trigones, and the cuticular verruculae are distinct.

The material examined by the writer shows a number of female branches with unfertilized archegonia and a very few with perianths. The perichaetial leaves are in three or four series and show the usual increase in size in passing from the base upward. Those of the innermost series (Fig. 36, H) measure 1.3—1.7 mm. in length by 0.35—0.5 mm. in width. They are narrowly ovate in outline and are bifid to about the middle with slender tapering divisions, each of which is tipped with a row of two to five long cells (Fig. 36, K). The margins are variously dentate to ciliate (Fig. 36, J), and the cilia are, in many cases, curved or contorted. The leaves of the next outer series (Fig. 36, I) are a little shorter than those of the innermost series, but are otherwise similar to them. The leaves of the outermost series are still smaller, measuring only 0.45 mm. in length by 0.3 mm. in width, and the margins of these leaves are sparingly denticulate. The cells of the perichaetial leaves are thin-walled, without trigones, and the cuticle is very faintly striolate or striolate-verruculose.

The perianth is narrowly ovate in outline and measures 2—2.5 mm. in length by about 0.5 mm. in diameter. It conforms to type by being terete below and tricarinate above, with rounded keels separated by deep grooves. The contracted and plicate mouth is deeply laciniate, and the laciniae are numerous or at least more than five. In all probability, however, their number is indefinite. The laciniae, which measure 0.4—0.7 mm. in length, are capillary or narrowly subulate, in the latter case tapering from a base two to four cells wide to a capillary apex consisting of several long cells in a row (Fig. 36, *L*). The sides of the subulate laciniae are irregularly dentate from projecting cells or short-ciliate. The cells of the perianth below the laciniae are 40—70 μ in length by 15—20 μ in width; those in the laciniae are about as long but only 10 μ wide. The walls and cuticle are essentially the same as in the cells of the perichaetial leaves.

Excellent figures of *A. inaequilaterum* are to be found in LINDENBERG and GOTTSCHE's monograph (20, *pl. 4, f. 1—5*). They were drawn from WALLICH's original Nepal material, and the same thing is true of the figures in STEPHANI's unpublished Icones. The writer also has had the privilege of studying some of WALLICH's specimens, which presumably represent the species in its most typical development. The leaves in these specimens show clearly the subrectangular form (Fig. 33, *B*) assigned to the leaves in the foregoing description, and the margin is closely crenulate or denticulate throughout the greater part of its extent. Nearly all the teeth represent nothing more than slight projections of marginal cells (Fig. 33, *C*), and even the teeth which project farther than the others are still unicellular. On both surfaces of the leaf, at least in the apical portion, the cell-walls form low, rounded or bluntly pointed protuberances (Fig. 33, *F*). The projecting walls are, for the most part, uniformly thickened. In some cells these walls are thicker in the middle than elsewhere, but the thickness gradually diminishes toward the periphery, and it is only in rare instances that an actual tubercle is developed. A few of the underleaves develop a distinctly larger tooth on one side or even on each side, but the majority fail to develop such teeth.

Two types of variation from the normal condition of *A. inaequilaterum*, as exemplified by WALLICH's specimens, can be distinguished. In the first type the marginal teeth are more irregular, and one or more multicellular teeth are present on some of the leaves. It was upon plants of this type that DE NOTARIS based his *Mastigobryum elegantulum* from Borneo. STEPHANI (36, p. 243) changed the name of this plant to *M. Notarisii* because there was an older *M. elegantulum* Gottsche from the island of Jamaica. The writer has studied BECCARI's original material of *M. elegantulum* and finds that it agrees in most respects with the description and figures published by DE NOTARIS. There is one important point, however, in which

the difference between *M. elegantulum* and typical *A. inaequila-*
terum is less than DE NOTARIS implies. This difference concerns the
leaf-cells. According to the description the cells of *M. elegantulum*
are plane on the surface, and DE NOTARIS states that this difference
by itself would be sufficient to distinguish the species from *A. inae-*
quilaterum. As a matter of fact the type-material of *M. elegantulum*
shows that the cells are essentially like those of WALLICH's specimens.
In other words the free cell-walls project as low protuberances,
which may be thicker in the middle than at the periphery; and in
rare instances, just as in the Nepal specimens, tubercles are present.
Similar tubercles, moreover, can occasionally be demonstrated on
the underleaves and on the leaves of the flagelliform branches.

With the elimination of the supposed difference in the leaf-cells,
the leaves of *M. elegantulum* still differ from those of WALLICH's
specimens in their more diverse teeth, some of which are multicellular. If leaves of these two plants are compared (Fig. 33, *B*, with
Fig. 34, *A*) the difference is indeed striking and might, at first sight,
seem sufficient to warrant a specific separation. The difference is
bridged over, however, by some of the other specimens listed. In
BARTLETT's No. 7075 from Sumatra, for example, the diversity in
the size of the teeth (Fig. 32, *A*, *B*) is less pronounced than in the
Bornean plants but more pronounced than in the Nepal plants.
Multicellular teeth, moreover, although relatively rare, can occasionally be demonstrated in these plants. Such intermediate
forms clearly break down the second distinction between *M. elegant-*
ulum and typical *A. inaequilaterum*.

The other type of variation illustrated by *A. inaequilaterum* has
to do with the development of tubercles on the leaf-cells. It has been
shown that tubercles are of rare occurrence in the Nepal specimens.
In some of the other specimens listed, however, the tubercles are
not only abundant but conspicuous. This is true, for example, of
VERDOORN's No. 79 from Mt. Ophir, of RIDLEY's No. 723 from the
same locality, of VERDOORN's No. 12 from Singapore, of some of
VAN DIEST's specimens from Banka, and of TEYSMANN's specimens
from Amboina. Plants of this character have sometimes made it
difficult to draw a sharp line between *A. inaequilaterum* and *Masti-*
gobryum echinatum and have served as the basis for some of the
records of the latter species in the literature. It may be noted in

this connection that STEPHANI's figures of *"M. echinatum"* in the
Icones were drawn from VAN DIEST's Banka specimens, which are
included in the present article under *A. inaequilaterum.*

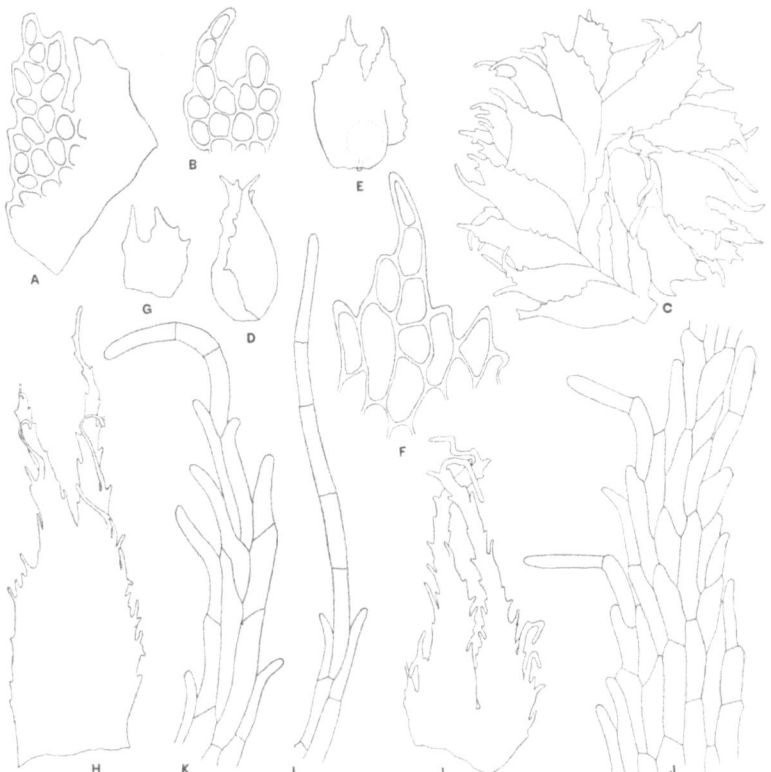

FIG. 36. *Acromastigum inaequilaterum* (Lehm. & Lindenb.) Evans. *A, B.*
Leaves of flagelliform branches, × 225. *C.* Male inflorescence, × 50. *D.* Male
bract, natural position, × 50. *E.* Male bract, spread out, × 50. *F.* Apex of
division of male bract, × 225. *G.* Male bracteole, × 50. *H.* Perichaetial leaf
of innermost series, × 40. *I.* Perichaetial leaf of second series, × 50. *J.* Part
of division of perichaetial leaf, innermost series, × 225. *K.* Apex of division
of perichaetial leaf, innermost series, × 225. *L.* Apex of lacinia at mouth of
perianth, × 225. The figures were drawn from specimens collected by
BARTLETT near the Aek Kanopan, Sumatra, Nos. 7075 and 7077.

VERDOORN's No. 79 differs from the Nepal specimens of *A. inae-*
quilaterum not only in its well-developed and abundant tubercles
(Fig. 35, *F, I*) but also in having somewhat larger leaf-cells with

thicker walls (Fig. 35, *C*). There are, therefore, rather marked differences between the two plants. Other specimens, however, with abundant tubercles have leaf-cells which are otherwise essentially like those of the Nepal specimens and still others have leaf-cells of an intermediate size (Fig. 35, *D*). In a few specimens, moreover the measurements of the leaf-cells vary considerably in different parts of an individual plant. BUCH (3, p. 14, 15) has shown that in the genus *Scapania* the size of the leaf-cells, the deposition of secondary layers of thickening on the cell-walls, and the development of cuticular structures are all modified to a greater or less extent by differences in the environmental factors. In all probability the genus *Acromastigum* is similarly susceptible, and the writer is inclined, at least for the present, to interpret the tuberculate plants of *A. inae-quilaterum* as a modification, in BUCH's sense, rather than as a taxonomic form or variety. Of course further investigations may show that this opinion is untenable.

So far as the marginal teeth of the leaves are concerned most of the tuberculate plants are in essential agreement with the Nepal specimens, except that crenulations predominate over denticulations. In a few of the plants, however, larger teeth are present on some of the leaves, and these larger teeth, in rare instances, may even be multicellular. It may be noted, also, that distinct lateral teeth (Fig. 35, *J*, *K*) are not uncommon on the underleaves, but they are far from being a constant feature.

25. **Acromastigum bifidum** (Steph.) comb. nov.

Mastigobryum bifidum Steph. Spec. Hepat. 3: 537. 1909.

The species is known only from the type-material, which was collected on an island east of New Guinea. This may be recorded as follows:

D'Entrecastaux Archipelago: Fergusson Island, without date, W. MICHOLITZ (H.), type of *Mastigobryum bifidum*.

The material studied by the writer is pale greenish brown and indicates that the plants grew in depressed mats. The living portions were 0.5—1 cm.in length, and the apparent dichotomies occur at intervals of 2—5 mm. The lateral leafy branches, which conform to the *Frullania* type, are supplemented by ventral leafy branches, which are intercalary in origin, arising in the axils of underleaves. Branches of this character have already been noted in *A. inaequilaterum*, but are of more frequent occurrence in *A. bifidum*. The leaves of the ventral branches, at least toward the base, are smaller than those of the lateral branches. The flagelliform branches, just as in *A. inaequilaterum*, conform to the *Acromastigum* type. Well-developed dorsiventral branches are 0.12—0.18 mm. wide and 0.11—0.17 mm. thick. The cortical cells (Fig. 37, *A*) are 35—50 μ in tangential width by 30—40 μ in radial width, and the medullary cells have an average diameter of about 20 μ. The outer walls of the cortical cells may be as much as 10 μ in thickness, but the walls of the medullary cells are thin, except for the triangular thickenings at the angles.

The leaves are loosely imbricated and lie approximately in a single plane, although the upper surface is a little convex. They spread at an angle of 60 to 80 degrees or even, in exceptional cases, at an angle of 90 degrees. Well-developed examples are 0.45—0.6 mm. in length by 0.18—0.24 mm. in width and show an unsymmetrically ovate-oblong from (Fig. 37, *B*) with little or no curvature. The leaves are rounded at the dorsal base and arch to about the middle of the branch. Beyond the base the dorsal margin extends as an almost straight line to the apex of the dorsal division, which usually points directly outward. The ventral margin is subparallel with the dorsal to about the bottom of the sinus and then curves or bends forward to the apex of the ventral division, which usually points obliquely forward. The narrow and acute sinus is one-fourth to one-half the length of the leaf, and the divisions, which are always close together, may even overlap a little. These divisions are both triangular and taper gradually to sharp points, which are tipped with a single cell or with a row of two cells. The length of the ventral division (Fig. 37, *B*, *F*) is a little more than that of the dorsal division (Fig. 37, *E*), but the width at the base varies from a little less to a little more. In a series of leaves examined the dorsal divisions were six to eight cells wide and the ventral divisions five or six cells wide. The leaf-margins are minutely crenulate or denticulate throughout from projecting cells. In some cases every cell projects over a considerable distance, and the teeth in consequence are close together. In other cases, particularly along the ventral margin, some of the cells do not project at all, and the teeth are then more scattered and irregular. Multicellular teeth are of very rare occurrence.

The vitta (Fig. 37, C), although passing insensibly into the adjacent tissues, can be distinguished from the base of the leaf to about the middle. It is two to four cells wide and is separated from the ventral

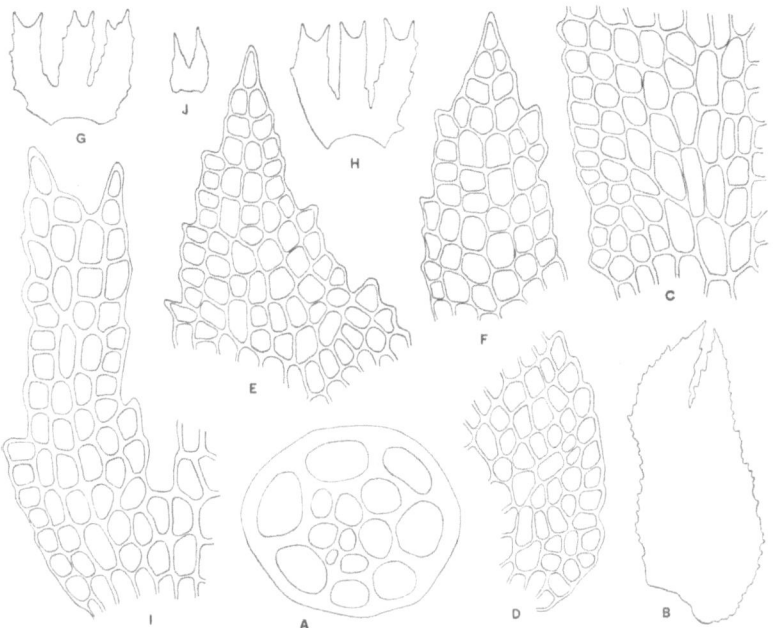

FIG. 37. *Acromastigum bifidum* (Steph.) Evans. *A*. Cross-section of branch, × 225. *B*. Leaf, × 50. *C*. Cells from base of a second leaf, ventral side, × 225. *D*. Dorsal base of second leaf, × 225. *E*. Dorsal division of second leaf, × 225. *F*. Ventral division of second leaf, × 225. *G, H*. Underleaves, × 50. *I*. Lateral division of a third underleaf, × 225. *J*. Leaf of flagelliform branch, × 50. The figures were drawn from the type-material.

margin by two or three rows of cells and from the dorsal by ten to sixteen rows. The cells of the vitta measure 20—40 μ in length by 15—20 μ in width. The cells in the vicinity of the dorsal margin (Fig. 37, *D*) have an average diameter of about 11 μ, and those in the dorsal division measure about 20 × 15 μ. The walls are everywhere uniformly thickened and attain a thickness of 3 μ in the vitta and vicinity. Neither trigones nor pits are apparant. The superficial walls bulge on the dorsal leaf-surface and also, in many cases, on the ventral surface as well, in the apical part of the leaf; and the bulging portion may show a minute median tubercle. The cuticle is minutely verruculose, but the verruculae are difficult to demonstrate, except along the margin and on the tubercles.

The underleaves are loosely imbricated and spread a little at the base, but the greater part of their surface lies subparallel with the surface of the branch. They have a subquadrate form (Fig. 37, *G, H*), narrowing slightly toward the base, and the sides are straight or a little convex. Well-developed examples are 0.2—0.25 mm. long. The divisions are subparallel in their natural position but diverge slightly when dissected off and spread out flat. The narrow sinuses, which are acute to rounded at the bottom, are two-thirds to three-fourths the length of the underleaf. The divisions are ligulate (Fig. 37, *I*), four to six cells wide at the base and, in most cases, four cells wide in the vicinity of the apex. The latter varies from retuse to distinctly bidentate, and the sharp teeth are either unicellular or consist of a row of two cells. The basal part of the underleaf is three to five cells high. The margins of the basal part and of the divisions are irregularly crenulate or denticulate. An occasional tooth may be sharper and larger than its neighbors, but no multicellular teeth have been observed.

The scale-like leaves of the flagelliform branches are ovate (Fig. 37, *J*) and the best developed examples measure about 0.15 mm. in length by 0.07 mm. in width. They are bifid to below the middle, with slender, subparallel divisions, the apices of which are normally acuminate and tipped with a row of two to four cells. In some instances, however, the divisions are shorter and bidenticulate at the apex. The margins of the leaves bear a few scattered crenulations or denticulations. No sexual branches are present in the material.

The relationship between *A. inaequilaterum* and *A. bifidum* is very close, and the attempt to maintain the validity of the latter is perhaps open to question. The two species agree in color, in size, in the general form of their leaves and underleaves, and in most of the characters derived from the leaf-cells and from the marginal crenulations and denticulations. There are, however, a few points of difference; and, if these can be proved constant through the study of more material, the claims of *A. bifidum* for recognition will have a firmer basis. The distinctions between the species have to do mostly with the depth of the sinuses in the leaves and underleaves and with slight differences in the form of the leaf- and underleaf-divisions. In *A. inaequilaterum*, for example, the sinus in the leaves is relatively short and rarely exceeds one-fourth the length of the leaf, the divisions also are thus relatively short, the ventral division is invariably wider than the dorsal, the sinuses of the underleaves scarcely extend beyond the middle, and the divisions thus scarcely exceed the basal portion in length and present a somewhat squat appearance. In *A. bifidum*, on the other hand, the sinus in the leaves is one-fourth to one-half the length of the leaf, the divisions are thus relatively longer than in *A. inaequilaterum*, the ventral division may be narrower than the dorsal, the sinuses of the underleaves extend two-thirds to three-fourths the distance from apex to base, and the divisions are thus distinctly longer than the basal portion and present an elongate, ligulate appearance.

There is also a marked resemblance between *A. bifidum* and *A. denticulatum*, although the plane superficial walls in the leaf-cells of the latter species will at once serve to distinguish it. The marginal crenulations and denticulations in *A. denticulatum*, moreover, although more numerous than in *A. linganum*, are less numerous than in *A. bifidum*, and many of the marginal cells fail to project at all.

26. **Acromastigum echinatum** (Gottsche) comb. nov.

Jungermannia inaequilatera β. *minor* Lehm. & Lindenb. in Lehmann, Pug. Plant. 6: 56. 1834.
Mastigobryum echinatum Gottsche in G. L. & N. Syn. Hepat. 218. 1845.
Jungermannia dirhyncha Tayl. ms., in part, *l.c.*, as synonym of *M. echinatum*.
Bazzania echinata Trevis. Mem. Ist. Lomb. 13: 414. 1877.

The type-material of the present species, as already noted under *A. inaequilaterum*, was collected by WALLICH at Singapore. This was the only station definitely known to GOTTSCHE, although he doubtfully listed Tasmania as a second station, since the synonymous *Junger-mannia dirhyncha* Tayl. was said to have come from that island. No mention is made of Tasmania, however, in connection with *M. echinatum*, in LINDENBERG & GOTTSCHE's monograph (20), so that the original report was evidently an error. Since the publication of the monograph the species has been reported from a number of Indo-Malayan stations, in addition to Singapore. Some of these reports, in the opinion of the writer, were based on specimens of *A. inaequila-terum* and have already been referred to under that species; the others, which are few in number, will be noted below. The following specimens of the true *A. echinatum* have been examined:

B a n k a: between Kimat and Simpang, 1858, S. KURZ (L.); between Klappa and Tjantara, 1858, S. KURZ (L.). Both of these stations are listed by VAN DER SANDE LACOSTE (28, p. 301).

B o r n e o: near Martapoera, Dutch Borneo, without date, P. W. KORTHALS (L.). VAN DER SANDE LACOSTE does not cite this station, although the specimens here listed are in his herbarium. He does, however, cite the species from Sumatra (28, p. 301) on the basis of specimens collected by KORTHALS. These specimens are apparently not in his herbarium. It is possible, therefore, that his record for Sumatra was based on the specimens from Borneo.

M a l a c c a: Singapore, without date, N. WALLICH (B., N. Y.), type of *Jungermannia inaequilatera* β. *minor* Lehm. & Lindenb. and of *Mastigobryum echinatum* Gottsche; same locality, DOWN (N. Y.); Mt. Ophir, Johore, without date or collector's name (N. Y.), specimen from the Girffith Herbarium; same locality, 1930, F. VERDOORN 48 (V., Y.).

STEPHANI (39, p. 536) records *A. echinatum* from Amboina, Banca, Borneo, Malacca, Great Natuna Island, Malacca, Singapore, and Sumatra, without giving the names of the collectors or citing more definite localities. Most of the specimens upon which his records were based have probably been referred to in the preceding pages. According to HERZOG (16, pp. 201 and 341) the species is a widely distrib-uted Indo-Malayan endemic.

The general features of *A. echinatum* are clearly shown by GOTT-SCHE and LINDENBERG (20, *pl. 4, f. 6—10*). The species is much like *A. inaequilaterum* in habit and in color, although a distinct yellowish pigmentation of the cell-walls is apparent in most cases. The living portions of the plants are about 1 cm. in length and the dichotomies occur at intervals of 2—3 mm. The ordinary leafy branches are 0.14—0.17 mm. in width by 0.12—0.15 mm. in thickness, and the pigmentation of the cell-walls shows with especial

FIG. 38. *Acromastigum echinatum* (Gottsche) Evans. *A*. Cross-section of branch, × 225. *B*, *C*. Leaves, × 50. *D*. Cells from base of leaf shown in *B*, ventral side, × 225. *E*. Dorsal division of same leaf, × 225. *F*. Ventral division of same leaf, × 225. *G*. Cross-section of leaf, × 300. *H*, *I*. Underleaves, × 50. *J*. Lateral division of underleaf shown in *H*, × 225. *A* and *G* were drawn from specimens collected by DOWN at Singapore; the remaining figures from the type-material, collected by WALLICH in the same locality.

clearness in cross-sections. The cortical cells (Fig. 38, *A*) measure 30—60 μ in tangential width by 30—40 μ in radial width, and the medullary cells average about 20 μ in diameter. They thus agree closely with the corresponding cells of *A. inaequilaterum* in size. The outermost walls of the cortical cells are, in most cases, 10—12 μ thick;

the walls separating the medullary cells are much thinner but show distinct thickenings at the angles.

The leaves are loosely to closely imbricated and spread at an angle of about 90 degrees. The upper surface is more or less convex and the outer part may be deflexed; in many cases, however, the leaves on both sides of the axis lie approximately in a single plane. Well-developed leaves measure 0.5—0.6 mm. in length by 0.25—0.35 mm. in width. They may appear slightly falcate, when dissected off and spread out flat (Fig. 38, *B*, *C*), but this peculiarity is scarcely noticeable in attached leaves. Although they show an approach to a rectangular form, this is much less evident than in *A. inaequilaterum*, and the leaves might better be described as ovate or oblong-ovate. The dorsal base is rounded, and the leaves arch slightly beyond the middle of the axis. Except at the base the dorsal margin is straight or slightly convex throughout its entire extent. The ventral margin is straight or slightly concave for about two-thirds its length and then curves or bends gently forward. The sinus, which is acute to rounded, is one-fifth to one-fourth the length of the leaf, if measured from the apex of the ventral division to the base, and the angle formed by its sides varies between 45 and 90 degrees. The ventral division (Fig. 38, *F*) is ovate or oblong-ovate and, in some cases at least, curves forward, although the curvature is never so great as in typical *A. inaequilaterum*. The dorsal division (Fig. 38, *E*) which is always shorter than the ventral and which may be greatly reduced in size, is triangular and points directly outward. Both divisions are acute to rounded, and a single terminal cell is distinguishable in some cases but not in all. The margin of the leaf is minutely and irregularly crenulate from projecting cells, but the normal crenulations are often obscured by the presence of cuticular outgrowths of various sorts. These will be described below.

The vitta is two or three cells wide and is separated from the ventral margin by two or three rows of cells. It is even more vaguely defined than in *A. inaequilaterum*, and its cells are distinguished by their arrangement in longitudinal rows, rather than by their elongate form (Fig. 38, *D*). These cells, for the most part, are 20—30 μ in length by 20—25 μ in width. In the other parts of the leaf the cells still average 20—25 μ in diameter, except along the margin, where the average width is about 15 μ. Trigones are everywhere conspicuous and are separated by distinct pits. The sides of the trigones are usually convex but may be straight. The coalescence of two trigones is a frequent phenomenon, and the coalescences in many cases show a definite rectangular or rhombic form. Except at the base of the leaf, each cell bears a tubercle 8—12 μ in diameter and in height on its dorsal surface (Fig. 38, F, G). This tubercle represents a local thickening of the free convex cell-wall and is in the form of a short cylinder or truncated cone width a broad rounded apex. In most of the cells the tubercles are restricted to the dorsal surface; in the vicinity of the apex, however, some of the cells bear tubercles on the ventral surface as well. Some of the marginal cells in this region may, in fact, develop a third tubercle on the free marginal wall. The surface of the tubercles is covered over with rather coarse verruculae, and similar verruculae are often found in the depressions between the tubercles. These various outgrowths give the leaf a frosted appearance under the microscope. The verruculae and larger cuticular outgrowths are especially conspicuous along the margin of the leaf. The tubercles here may lie opposite the cavities of the marginal cells, as they normally should, but in some cases low tubercles alternate with the marginal cells and thus lie opposite the walls between them (Fig. 38, D).

The underleaves are loosely to closely imbricated and distinctly convex, when seen from below, without being appressed to the axis. They measure, when well developed, 0.22—0.28 mm. in length by 0.3—0.35 mm. in width and are broadly subrectangular, when dissected off and spread out flat (Fig. 38, H, I). The subequal divisions are separated by narrow acute sinuses, which extend a little beyond the middle. The divisions are ligulate-ovate and taper slightly to a retuse, truncate, or subacute apex. In most cases they are four cells wide at the base and two cells wide at the apex (Fig. 38, J). The base may, however, be five cells wide, and the apex may be tipped with a single cell. The wall-thickenings in the cells of the underleaves are essentially like those in the leaf-cells. The verrucae, however, are restricted to the ventral walls, i.e., to the walls turned away from the axis. The marginal crenulations formed by projecting cells are few and far between, and yet the presence of verruculae and occasional verrucae give the margin an irregular appearance.

The flagelliform branches are 0.9—0.12 in diameter. Their ovate leaves, which may attain a length of 0.15 mm. and a width of 0.12 mm., are bifid to about the middle with an acute to rounded sinus and acute or vaguely retuse divisions. The cells of these leaves show distinct trigones and verrucae, and the latter (as in the underleaves) are restricted to the surface turned away from the axis.

The material of *A. echinatum* at the writer's disposal is sterile, but GOTTSCHE and LINDENBERG have given a brief description of the female inflorescence and have represented a female branch with perianth on their plate (20, p. 17, *pl. 4, f. 10*). According to their account the perichaetial leaves are bifid to the middle, with lanceolate subulate divisions, and the lateral margins may be angular-dentate or bear a single dentiform-subulate lacinia. The perianth is said to be plurifid at the mouth, with long lanceolate-subulate laciniae. The male inflorescence of *A. echinatum* is apparently still unknown.

The description of *A. echinatum* brings out many points of agreement between the species and *A. inaequilaterum*. The plants of both are of about the same size, the cortical and medullary cells of the axes have important features in common, the leaves are similar in form, and the leaf-margins are minutely toothed. Of course specimens of *A. inaequilaterum* in which the leaf-cells are desitute of surface-tubercles or in which the tubercles are only slightly developed can be distinguished at a glance from *A. echinatum*, in which the tubercles are both abundant and conspicuous. It has been shown, however, that specimens of *A. inaequilaterum* sometimes occur in which the tubercles are just as abundant as in *A. echinatum* and are also of a fairly large size. In specimens like these the most trustworthy distinctions are those drawn from the thickenings of the vertical walls of the leaf-cells. In *A. inaequilaterum* these thickenings are usually uniform; and, even when vague indications of trigones can be demonstrated, the pits between them are filled with deposits of cell-wall substance. In *A. echinatum*, on the other hand, the thickenings are in the form of definite trigones, and the pits between them are both persistent and conspicuous. Of course the coalescence of two trigones, theoretically at least, involves the obliteration of a pit. This phenomenon is of frequent occurrence in *A. echinatum*, and the coalescence of more than two trigones can occasionally be demonstrated. But, even in such cases, the number of pits obliterated is relatively small, and the majority remain distinct. Coalescences are apparently more frequent and more extensive in the underleaves than in the leaves.

Other distinctions between *A. echinatum* and *A. inaequilaterum* can be found in the size of the leaf-cells and in the character of the marginal teeth of the leaves and underleaves. In *A. echinatum* the leaf-cells are a trifle larger than those of *A. inaequilaterum*, even when tuberculate specimens of the latter are considered; and the marginal teeth are, in general, less distinct. With rare exceptions the marginal teeth of *A. echinatum* are crenulations, rather than denticulations, many of the marginal cells do not project at all, and multicellular teeth are apparently never produced. In *A. inaequilaterum*, on the contrary, the marginal teeth are frequently denticulations, most of the marginal cells project as teeth, and multicellular teeth can be demonstrated on some of the leaves. In *A. echinatum*, moreover, marginal tubercles are often present and simulate the crenulations formed by projecting cells. These tubercles, together with smaller verruculae, give the margin a more ragged contour than in the leaves of *A. inaequilaterum*.

27. **Acromastigum fimbriatum** (Steph.) comb. nov.

Mastigobryum fimbriatum Steph. Spec. Hepat. 3: 538. 1909.

This remarkable species was based on specimens collected by EVERETT at Baram, Borneo. The data given in STEPHANI's Icones add that this specimen formed a part of No. 17. The MITTEN Herbarium contains an unnamed specimen from EVERETT's Baram collection, which forms a part of No. 9. Since this specimen agrees with STEPHANI's description and figures of *Mastigobryum fimbriatum* it may be considered authentic and may be recorded as follows:

B o r n e o: Baram, Sarawak, without date, A. H. EVERETT 9, in part, mixed with fragments of a *Trichocolea* (N. Y.), probable type of *Mastigobryum fimbriatum* Steph. No other specimens are definitely known to the writer, and HERZOG (16, p. 350) lists the species as endemic to Borneo.

The material studied by the writer is very fragmentary but indicates that *A. fimbriatum* is amply distinct from the other species of the genus. The plants when dry have a dingy yellowish brown color.

The living portions were about 1 cm. long, and the dichotomies occur at intervals of about 5 mm. Well-developed branches are about 0.15 mm. in width and 0.12 mm. in thickness. The cortical cells, as seen in cross-section (Fig. 39, *A*) have a tangential width of 30—70 μ and a radial width of 20—50 μ. The larger cortical cells are in the dorsi-lateral rows and the smaller in the ventral rows. The

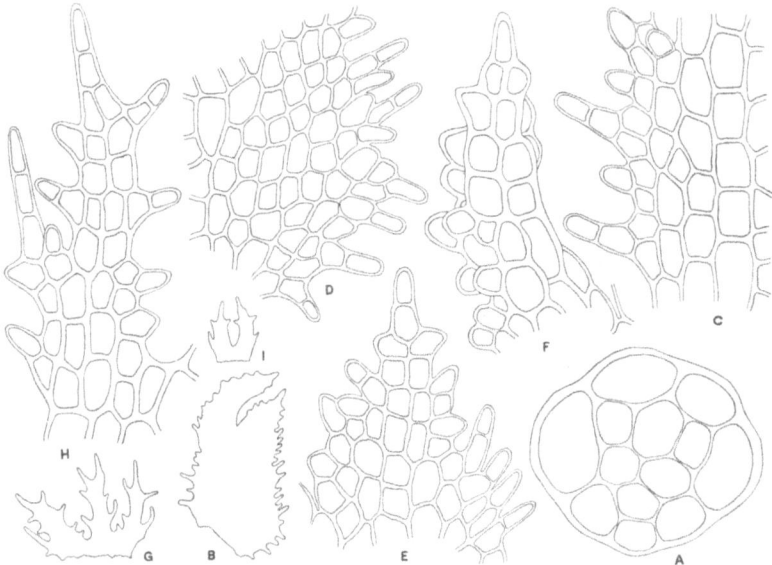

FIG. 39. *Acromastigum fimbriatum* (Steph.) Evans. *A*. Cross-section of branch, × 225. *B*. Leaf, × 50. *C*. Cells from base of same leaf, ventral side, × 225. *D*. Dorsal base of same leaf, × 225. *E*. Dorsal division of same leaf, × 225. *F*. Ventral division of same leaf, × 225. *G*. Underleaf, × 50. *H*. Lateral division of same underleaf, × 225. *I*. Leaf of flagelliform branch, × 50. The figures were drawn from specimens collected by EVERETT at Baram, Borneo, No. 9, in part.

latter are of about the same size in cross-section as the medullary cells, which average about 25 μ in diameter. The outer walls of the cortical cells are 4—6 μ thick; the medullary cells have much thinner walls, but show small thickenings at the angles. All the walls are faintly pigmented with yellowish.

The leaves spread at an angle of about 90 degrees and are loosely imbricated. They are not deflexed but lie approximately in a single plane. When a branch is examined from above the ventral portion of each leaf shows a shallow longitudinal groove extending into the ventral division, whereas the dorsal portion, including the dorsal division, is plane or nearly so. When the branch is examined from below the groove appears as a low rounded ridge. Well-developed leaves are about 0. 45 mm. long and 0.25 mm. wide. When dissected off and spread out as flat as possible they show an unsymmetrically oblong outline (Fig. 39, B), with the dorsal and ventral margins subparallel as far as the bottom of the sinus. The dorsal base is rounded (Fig. 39, D) but reaches scarcely beyond the middle of the axis. Beyond the base the margin extends as an almost straight line to the apex of the dorsal division, which points outward or slightly forward. The ventral margin, similarly, is almost straight as far as the bottom of the sinus and then curves forward, thus determining the direction of the ventral division, which points distinctly forward. The sinus, which is about one-third the length of the leaf, is narrow but forms a definite angle between the divisions, both of which are acute and tipped with a single cell or with a row of two cells. The dorsal division (Fig. 39, E) is triangular and tapers from a base five to seven cells wide. The ventral division (Fig. 39, F), which also is triangular, is a little longer than the dorsal division but also a little narrower, being only three to five cells wide at the base. The ridge on the ventral surface can not be flattened out by pressure.

The margins of the leaves are denticulate to short-ciliate throughout their entire extent, and each tooth consists of a single cell or of a row of two cells (Fig. 39, E, C). The teeth as a rule are closely crowded but are separated, in some cases, by one or (rarely) two cells which do not project. On the ventral division the teeth are mostly unicellular and point obliquely forward. They are supplemented on the convex ventral surface by superficial outgrowths, which consist of single cells (Fig. 39, F), of pairs of cells, or of short longitudinal ridges three or four cells long. One or two similar outgrowths on the ventral part of the leaf below the division can likewise be demonstrated on some of the leaves (Fig. 39, C, upper part on the left).

The vitta (Fig. 39, C), which is two or three cells wide at the base, is fairly distinct and can be followed into the ventral division. Toward the base of the leaf it is separated from the ventral margin by two or three rows of cells and from the dorsal margin by five to seven rows. Its cells are mostly 25—40 μ long and 20—25 μ wide. The cells near the dorsal margin (Fig. 39, D) average about 18 μ in diameter and those along the margin itself about 20 μ. The cell-walls may attain a thickness of 4 μ in the vitta and vicinity but are only 2—3 μ thick elsewhere. The thickening is uniform throughout, without trigones or evident pits, and the superficial walls, which bulge slightly on one or both surfaces, are about as thick as the vertical walls and show no indications of median tubercles. The cuticle is smooth or nearly so throughout.

The loosely imbricated underleaves are somewhat convex on the ventral surface. They are trapezoidal in general outline (Fig. 39, G), broadening out somewhat from the base, and measure about 0.3 mm. in length by 0.35 mm. in greatest width. The divisions are separated by acute to lunulate sinuses, from two-thirds to three-fourths the length of the underleaf, and the lateral divisions may diverge slightly from the median. The divisions, which do not vary greatly in length, are subulate in form, two to four cells wide at the base, and tipped with a row of two to four cells. The margins (Fig. 39, H) are dentate to short-ciliate, but the teeth tend to be less crowded and longer than on the leaves. An occasional cilium, in fact, may be three or four cells long and may even bear secondary denticulations (Fig. 39, H, at left). The cells of the underleaves are essentially like the leaf-cells.

Some of the flagelliform branches have a diameter of 0.15 mm. and are therefore no more slender than the ordinary dorsiventral branches. The scattered scale-like leaves have an ovate form (Fig. 39, I) and may be as much as 0.15 mm. in length. They resemble the underleaves to a certain extent but are bifid instead of trifid and bear fewer and shorter marginal teeth.

The leaves of A. fimbriatum present a number of unusual features, such as the well-developed marginal teeth, the ridge along the ventral side, and the superficial outgrowths on this ridge. The marginal teeth are better developed than in any of the other species that have so far been considered and give the leaves an almost ragged appearance. They are not so distinctive, however, as the ridge with its outgrowths. In all the other species the lower leaf-surface is concave along the ventral side, when a branch is examined from below, and in many cases the concavity is so great that the ventral margin appears in profile view, at least toward the base. The superficial outgrowths also are unique, so far as the genus *Acromastigum* is concerned, but find their homologues in the surface-hairs of *Lophocolea muricata* Nees (see HERZOG, 16, *f. 5, 76*). The outgrowths give rise to bistratose regions in the leaf, but these are different from the bistratose regions in the leaves of *A. brachyphyllum*. In the latter species the leaf-surfaces do not bulge perceptibly in such regions, but in *A. fimbriatum* the bistratose condition is associated with definite projections.

28. **Acromastigum lobuliferum** sp. nov.

Mediocre, flavo-viride, hymenophyllis et aliis hepaticis consociatum; caules parce ramosi; folia imbricata, oblique patula, decurva, ovato-oblonga, 0.45—0.6 mm. longa, 0.3—0.35 mm. lata, bifida, lobis subaequalibus vel ventrali parum longiore, triangulatis, normaliter acutis, margine irregulariter crenulato, denticulato et dentato, paucis dentibus ventralibus lobuliformibus; cellulae in parte ventrali circa 18 μ lata, in parte dorsali circa 12 μ lata, parietibus incrassatis, trigonis nullis; foliola imbricata, trifida, lobis breviter ligulatis, apice retusis vel bidentatis, margine denticulato, uno dente laterali magno; flores ignoti.

B o r n e o: Afd. Koetei, Boven Maratam Geb., 1898—99, AMDJAH 109, Exp. NIEUWENHUIS (Bog., Y.). Known only from the type-material.

The plants are scattered in a mat of filmy ferns and show a slight admixture with *A. laevigatum* and a few other hepatics. The color in dried material is a pale yellowish green, and the cell-walls are unpigmented throughout. The living portions of the stems are mostly 0.5—1 cm. in length, and the successive dichotomies are 3—5 mm. apart. Well-developed leafy branches are 0.15 mm. in width by 0.13—0.15 mm. in thickness. The cortical cells (Fig. 40, *B*) have a tangential width of 30—50 μ and a radial width of 25—40 μ, whereas the medullary cells average about 16 μ in diameter. The outermost walls of the cortical cells often attain a thickness of 10—14 μ; the walls separating the medullary cells are much thinner but show thickenings at the angles, and these thickenings often coalesce. The pits are not clearly defined.

The more or less closely imbricated leaves (Fig. 40, *A*) have a
distinctly convex upper surface except in the dorsal part, which is

FIG. 40. *Acromastigum lobuliferum* Evans. *A*. Part of plant, showing dorsiventral and flagelliform branches, ventral view, × 50. *B*. Cross-section of
branch with base of leaf, × 225. *C-E*. Leaves, × 50. *F*. Cells from base of
leaf, ventral side, × 225. *G*. Dorsal base of leaf, × 225. *H*. Ventral division
of leaf, × 225. *I*. Underleaf, × 50. *J*. Lateral division and tooth of underleaf,
× 225. *K*. Incomplete underleaf at base of flagelliform branch, × 50. *L*. Leaf
of flagelliform branch. The figures were drawn from the type-material.

plane or nearly so. In the ventral part the convexity is so marked
that the leaf is revolute except at the base, where the ventral margni

appears in profile view from below. The revolute portion usually involves the ventral division and the apical portion of the dorsal division, in consequence of which the apices of the divisions, in many cases, point obliquely inward and forward. The morphological features of the leaves cannot be clearly distinguished on intact branches; and even when the leaves are dissected off and spread out flat, they show so much diversity in form and in the character and distribution of the marginal teeth that it is somewhat difficult to select the more typical examples. The leaves shown in Fig. 40, C—E, are perhaps as nearly typical as any. Such leaves are ovate-oblong in form and measure 0.45—0.6 mm. in length by 0.3—0.35 mm. in width. At the dorsal base a slight rounded expansion is present, and beyond this the margin follows a straight or slightly convex line to the apex of the dorsal division. The ventral margin, taken in its entirety, is subparallel with the dorsal for about half its length and then bends or curves forward to the tip of the ventral division. The divisions exhibit considerable diversity with respect to their relative lengths. In some cases the ventral division is distinctly longer than the dorsal (Fig. 40, C); in other cases they are subequal in length (Fig. 40, D, E). In most of the leaves both divisions are triangular, acute, and tipped with a single cell (Fig. 40, H) or (more rarely) with a row of two cells; occasionally, however, obtuse, rounded or even truncate divisions are met with. The sinus, which is one-fourth to one-third as long as the leaf, is acute and narrow, and the divisions may be in contact with each other or even overlap. In a series of leaves examined the dorsal divisions were eight to ten cells wide at the base and the ventral divisions six to nine cells wide. The dorsal divisions thus show a tendency to be slightly wider than the ventral.

Except near the ventral base (Fig. 40, F) the margin of the leaves is closely and irregularly crenulate, denticulate, or dentate throughout. Along the dorsal margin (Fig. 40, G) nearly every cell projects. Most of the projections are in the form of crenulations or bluntly pointed denticulations, separated by shallow indentations. On many of the leaves, however, in addition to unicellular projections, one to four larger teeth are present, extending beyond the others and pointing obliquely forward. The largest of these teeth are perhaps three or four cells long and three or four cells wide at the base, but there are all intergradations between teeth of this character and the unicellular teeth. The larger teeth are usually acute and their margins, in some cases at least, are crenulate or denticulate. The ventral margin is even more irregular, with respect to its teeth, than the dorsal margin. At about the middle an unusually large tooth, or lobule, is present. This is directed forward and, on intact plants, is reflexed with the rest of the ventral margin. The lobule is five or six cells wide at the base and seven to ten cells long, the apex is usually acute, and the margin is crenulate, denticulate, or even sparingly dentate. Between the lobule and the apex of the ventral division one or two similar but shorter teeth are usually developed. Aside from the large teeth the ventral margin shows the usual unicellular crenulations and denticulations, and these are present also along the margins of the divisions. On some of the leaves the ventral division is no larger then the ventral lobes or teeth just described, and such leaves may appear to be irregularly trifid or quadrifid.

The arrangement of the leaf-cells in longitudinal rows is more or less apparent from the base to about the middle. In a band three or four cells wide, the cells tend to be larger than the other cells and longer than broad. This band (Fig. 40, F) may be interpreted as the vitta, although its boundaries are very vaguely defined. The vitta is separated from the ventral margin by two or three rows of cells and from the dorsal margin by twelve to fifteen rows. The cells of the vitta average about $18\,\mu$ in width and measure $18—30\,\mu$ in length; the cells toward the dorsal margin average about $12\,\mu$ in diameter and those in the divisions about $15\,\mu$. The walls between the cells are uniformly thickened, and neither pits nor trigones are apparent. In the vitta and vicinity the walls may be as much as $6\,\mu$ thick, the middle lamellae show with unusual distinctness, and the cell-cavities often retain their original polygonal outline (Fig. 40, F). Except in the basal part of the leaf each cell bears on its upper surface-wall a more or less distinct median tubercle. The tubercles are largest in the apical region and may be as much as $6—8\,\mu$ in height; toward the base they gradually become lower and less distinct and finally disappear. The surface of the leaf, aside from the tubercles is minutely and closely verruculose, but the verruculae are often difficult to demonstrate and may not be everywhere present. They are most distinct on the rounded summits of the tubercles.

The underleaves are imbricated and appear distinctly convex, especially in the basal part, when a branch is examined from below. When dissected off they vary in outline from subtrapeziform (Fig. 40, *I*), with the narrower parallel side at the base, to broadly subquadrate. Well-developed underleaves are 0.2—0.3 mm. long and 0.45—0.55 mm. wide, but many examples are considerably smaller. The three principal divisions are supplemented, in typical cases, by a large, spreading, lobe-like tooth on each side and are separated by narrow sinuses, which extend about halfway to the base. The sides of the divisions are subparallel or somewhat convergent, and the apex varies from retuse to sharply bidentate (Fig. 40, *J*). Most of the divisions are eight to ten cells long and six cells wide at the base, but occasional examples may be five, seven, or eight cells wide. The lateral lobe-like teeth, which are mostly subulate from a base three to six cells wide, are tipped with a row of two or three cells or, more rarely, with a single cell. On poorly developed plants the lateral teeth may be much smaller or even absent altogether. Aside from the lateral teeth the margins of the underleaves bear scattered and irregular crenulations, denticulations and larger teeth, except toward the entire base. The cells of the underleaves are similar to the leaf-cells and each cell, except in the basal region, bears a median tubercle on the outer, or ventral, surface. The incomplete underleaves at the base of the flagelliform branches, in the few cases examined, lacked a lateral tooth on the side turned away from the branch. In two such underleaves, however, one of which is illustrated (Fig. 40, *K*), a small basal tooth was present on the side toward the branch.

The flagelliform branches average about 0.1 mm. in diameter. Their convex scale-like leaves, which are approximate to contiguous toward the base of the branch, are subquadrate in outline (Fig. 40, *L*) and about 0.3 mm. long, when well developed. They are bifid to the middle or a little beyond, with an acute to lunulate sinus and slightly tapering divisions, the apices of which are typically bidentate. The divisions are much like those of the underleaves and may be as much as four to six cells wide at the base and ten cells long. On each side of the leaf at about the middle, a distinct tooth, three to five cells long and two to four cells wide at the base, is usually present. Marginal crenulations and denticulations also are present, and the characteristic tubercles, as in the underleaves, are restricted to the abaxial leaf-surface. In passing from the base of the branch outward the leaves undergo the usual decrease in size and simplification in structure.

The cuticular tubercles on the leaf-cells of *A. lobuliferum* indicate a relationship to *A. echinatum* and *A. inaequilaterum*. The tubercles are smaller than those of *A. echinatum* and are largely, if not entirely, restricted to the dorsal surface of the leaf. The small size of the tubercles is associated with the small size of the leaf-cells, which have an average diameter of only 15 μ in the leaf-divisions; in *A. echinatum* the cells in the same position have a diameter of 20—25 μ. The uniformly thickened walls of the leaf-cells agree with those of *A. inaequilaterum*, in showing neither trigones nor pits; the thickening, in fact, is unusually uniform in *A. lobuliferum* since the angles of the cell-cavities, in many cases at least, are not rounded.

The most remarkable feature of *A. lobuliferum* is the extreme irregularity exhibited by the marginal teeth of the leaves. Although many of the teeth, just as in *A. inaequilaterum*, are minute crenulations or denticulations, these minute teeth are associated with larger teeth which, in many cases, are multicellular. A somewhat similar irregularity has been noted in certain specimens of *A. inaequilaterum*, but the irregularity in *A. lobuliferum* is far more pronounced. Some of the teeth, for example, are large enough to be designated lobes or lobules, and there are all intergradations between these large and complex teeth and the minute teeth formed by projections of individual cells. The lobe-like teeth are usually produced singly on the ventral side of the leaves, and smaller but still multicellular teeth are frequently present between the lobe-like teeth and the apices of the ventral divisions. In *A. inaequilaterum* the multicellular teeth, in most cases, are restricted to the outer side of the ventral divisions. The two species thus show a similarity in the position of the larger teeth.

A tendency to produce large teeth is exhibited also by the underleaves and by the scale-like leaves of the flagelliform branches. In the case of the underleaves the large teeth are borne singly on the sides and diverge widely from the somewhat larger lateral divisions. A similar tendency to produce larger teeth on the sides of the underleaves is shown also by *A. inaequilaterum*, but the teeth in that species are smaller and less complex than those of *A. lobuliferum*. The teeth on the scale-like leaves of the latter species are smaller than those on the underleaves but are often distinct and occupy a similar lateral position.

PHYLOGENY AND AFFINITIES OF ACROMASTIGUM

The classification adopted in this revision proceeds in a general way from simpler and less highly differentiated to more complex and more highly differentiated forms. In *A. bancanum*, for example, which comes near the beginning of the series, the leaves and under-leaves are composed of a relatively small number of cells and are undivided. In *A. inaequilaterum*, on the other hand, which comes near the end of the series, the leaves and underleaves are composed of a relatively large number of cells, the leaves are definitely bifid, and the underleaves definitely trifid. A higher differentiation is further expressed by the marginal teeth of the leaves and under-leaves and by the frequent occurrence of cuticular tubercles.

The sequence adopted gives the student an orderly survey of the various species and also some idea of their relationship to one another. From the standpoint of phylogeny, however, the reverse sequence would perhaps have been more logical. In all probability the complex and highly differentiated species represent a more primitive condition than the more simply organized species. If this idea is correct the latter have been derived from the complex species by a reduction in size and by processes of coalescence, whereby parts which were originally separate have become congenitally fused. Indications of these processes are to be seen in several of the species and have already been noted in a number of cases, a few of which may now be reviewed. The leaves of *A. bidenticulatum*, for example, apparently represent a case in which the distinct leaf-divisions of the *Inaequilatera* have almost lost their identity and have become reduced to minute teeth. In this change both reduction in size and coalescence of the divisions seem to have been involved. The leaves of *A. bidenticulatum*, therefore, connect the bifid leaves of the *Inaequilatera* with the undivided leaves of the typical *Exilia*, in which the divisions have become coalescent throughout.

In *A. exiguum* a species is met with in which the process of reduction affects the dorsal leaf-divisions more particularly. It is possible
to arrange a series in this species leading from leaves with relatively
large dorsal divisions, through leaves with smaller and smaller dorsal
divisions, to leaves in which the dorsal division is reduced to a
unicellular tooth or is even obsolete. In this case the process of
coalescence is less in evidence than in *A. bidenticulatum*.

The xerophytic *A. tenax*, finally, represents a case in which the
divisions of the underleaves are involved. The trifid underleaves
constitute a distinctive feature of the more typical *Inaequilatera*,
but in *A. tenax* these divisions have almost disappeared in the
majority of the underleaves and have wholly disappeared in occasional examples. Here, as in the leaves of *A. bidenticulatum*, a
reduction in size has apparently been accompanied by a process of
coalescence.

If such forms as *A. inaequilaterum* are accepted as the more primitive types found in the genus the question regarding the origin
of these complex types naturally arises. Unfortunately the scanty
fossil remains of the Hepaticae throw little light on the ancestors of
the acrogynous liverworts. Most of the older fossils represent mar-
chantiaceous or anacrogynous types, which had already reached a
stage of differentiation approximating that of recent genera. In
determining what the ancestors of the present-day *Acrogynae* may
have been, the student has little to help him except the comparative
study of living types and certain phylogenetic hypotheses which
are more or less widely accepted. Some of these hypotheses have
to do with the sequence in time of various types of organization and
others with the doctrines of conservative structures and reversion.

It is generally assumed, for example, that erect, radial, and
isophyllous shoots precede, in phylogenetic lines, prostrate, dorsiventral, and anisophyllous shoots, and that the latter have been
directly derived from the former. GOEBEL has brought together
much evidence supporting this idea, particularly from his studies on
the mosses, the Lycopodiaceae, and certain groups of the phanero-
gams (13, p. 316—346). Another hypothesis affecting sequence is
that branch-systems showing a differentiation into various kinds of
branches have been derived from branch-systems in which the

branches were alike. Here again evidence drawn in part from the higher plants is available.

Two hypothetical sequences applying wholly or largely to the *Acrogynae* have been suggested by LEITGEB. According to the first, undivided leaves, as well as plurifid leaves, have been derived from bifid leaves (19, p. 11). The bifid type, therefore, represents the primitive leaf-type of the group. According to the second sequence, the terminal type of branching, in which the branches arise in very young segments, precedes the intercalary type, in which the branches arise in much older segments (19, p. 30). He presents evidence in support of both of these sequences.

As an example of a conservative structure in the leafy hepatics the female branch may be mentioned. In a number of highly differentiated genera the female branches differ from the ordinary vegetative branches and present features which have presumably been retained from less differentiated ancestors. Examples of reversion are afforded by the gemmiparous branches of certain species and by the flagelliform branches of others. Branches of this character also occur in highly differentiated genera and differ from the ordinary vegetative branches. In these cases the branches by a process of reversion, are supposed to approximate in certain features the vegetative branches of less differentiated ancestral forms.

If these various phylogenetic hypotheses, particularly those relating to sequences, are applied to the *Acrogynae*, some conception of the primitive representatives of the group may perhaps be gained. According to these hypotheses the ancestral types would have been erect plants, with three ranks of bifid leaves; their branches would have been terminal in origin and alike; and their ordinary vegetative leaves would have been essentially like the leaves associated with the sexual organs.

A survey of the various existing genera of the *Acrogynae* soon makes it evident that not a single one presents all of the primitive features enumerated. Perhaps *Herberta* approaches as close to this hypothetical condition as any. On the secondary stems of this genus, which are suberect, the leaves and underleaves are bifid and very much alike; the perigonial bracts and bracteoles, both of which bear antheridia in their axils (SCHIFFNER, 33), are essentially alike and differ only slightly from the foliage leaves; and the same thing is true of the perichaetial bracts and bracteoles. Even the three leaves which form the perianth are coalescent only in the lower part and the six laciniae at the mouth, which represent the free parts of the fused leaves, are strongly reminiscent of the divisions of ordinary leaves. In other respects, however, *Herberta* has advanced beyond the hypothetically primitive condition. The terminal branching, for example, has been entirely replaced by branching of the intercalary type, and the shoot-system shows a differentiation into a prostrate caudex with reduced leaves and the secondary stems described above.

Leafy hepatics which have advanced far beyond the primitive condition in certain respects may retain certain primitive features. The highly developed *Frullaniae*, for example, which show aniso-phylly clearly, accompanied by the presence of water-sacs on the leaves, still show only the terminal type of branching. In the genus *Anastrophyllum*, also, which has evolved in a different direction, the leaves have kept their primitive bifid form, although the anisophylly of the plants has brought with it the almost complete disappearance of the underleaves. The terminal type of branching, moreover, just as in *Herberta*, has been entirely supplanted by the intercalary type.

To illustrate the "conservatism" of the female branches such a species as *Cephalozia bicuspidata* (L.) Dumort. might be selected. The dorsiventral stems bear bifid leaves, but the underleaves are repre-sented merely by slime papillae and have thus undergone extreme reduction. On the sexual branches, however, three ranks of deeply bifid leaves are present; and the bracteoles, which represent the underleaves, are scarcely different from the bracts, which represent the leaves. The branches, therefore, which curve until suberect, deviate very slightly from the isophyllous branches of the ancestral types. The genus *Odontoschisma* is even more interesting in this respect. In the more characteristic species of this genus the vege-tative leaves are no longer bifid but undivided, and yet the three ranks of perichaetial leaves are bifid, just as in *Cephalozia bicuspi-data*, and therefore conform to the primitive leaf-type.

The gemmiparous branches of *Odontoschisma denudatum* (Mart.) Dumort. and the flagelliform branches of *Bazzania* will illustrate reversion. In the *Odontoschisma* the gemmiparous branches curve upward and assume an erect position. With the change in position of the branch the underleaves become larger and larger and the leaves smaller and smaller until, in the upper part of the branch, a condition of isophylly may be realized. In *Bazzania* the flagelliform branches grow downward and produce from the beginning three ranks of uniform scale-like leaves. Here again the branches are iso-phyllous, although the ordinary vegetative branches show aniso-phylly in a marked degree.

If the phylogenetic ideas brought forward are applied to *Acromas-tigum* it will be seen that the genus, on the whole, represents an advanced type of organization but that it shows at least two features of a primitive character. One of these is the presence of bifid leaves in the more complex species, and the other is the retention of terminal branching in both lateral and ventral segments. Two of the features which are advanced in character are the following : the dorsiventrality of the ordinary vegetative branches, accompanied by marked ani-sophylly; and the differentiation of the plant-body into dorsiventral vegetative branches, radial flagelliform branches, dorsiventral male branches and radial female branches. The relatively simple structure of the axial organs may likewise be regarded as evidence of reduction and therefore of advance, even if it represents the persistence of an early stage in ontogenetic development.

Although *Acromastigum* is closely related to *Bazzania*, there is one serious objection to STEPHANI's theory that the *Inaequilatera* repre-sent an old "Abzweigung vom normalen *Mastigobryum*-Typus" (39, p. 414). This is the occurrence of terminal branching in ventral segments. The sequence of the cell-walls laid down in young segments is, within certain limits, very stable, not only in individual species but also in certain genera and larger groups. This sequence leads, under ordinary circumstances, to the development of a leaf from each lateral segment and of an underleaf from each ventral segment. When a terminal branch is to be formed the normal sequence is modified in such a way that the apical cell of the branch becomes differentiated in one of the external cells of the young segment, leaving only a part of the segment for leaf-formation. A plant showing terminal branching in both lateral and ventral segments would therefore have two sequences at the disposal of each segment, one leading to normal leaf-formation and the other to the establishment of a terminal branch accompanied by an incomplete leaf. In the case of *Bazzania* it is only the lateral segments that have these two sequences at their disposal; the ventral segments have only the normal sequence, which leads invariably to the development of a complete underleaf. If *Acromastigum* represented a branch from the *Bazzania*-type it would have had to originate a second sequence in the early divisions of the ventral segments. This second sequence would have necessitated a profound change in the normal sequence, and the power of inducing such a change is, from an evolutionary standpoint, hardly conceivable.

It is much more probable that *Acromastigum* and *Bazzania* have descended from a common ancestral stock in which both lateral and ventral segments showed two distinct sequences in the early stages of cell-division, one leading to normal leaf-formation and the other to the development of a terminal branch accompanied by an incomplete leaf. In this hypothetical ancestral stock dorsiventrality with marked anisophylly and a differentiation of the plant body into various kinds of branches must have already been acquired. Intercalary branches in the axils of the underleaves, moreover, must have already appeared, and the development of the sexual organs must have been relegated to branches of this character. It is probable however, that the leaves of these ancestral plants still retained their bifid character and that the axial organs had a generalized type of structure. In such a type the tissues are only slightly differentiated, and the number of cells derived from each segment is indefinite and relatively large.

In diverging from this stock the genus *Acromastigum* retained terminal branching in the ventral segments and also bifid leaves, at least in the species which remained most primitive. The axial organs at the same time became differentiated into cortex and medulla, and the number of cell-divisions in the segments, leading to the development of the stem-tissues, became reduced. One result of this reduction is seen in the seven longitudinal rows of cortical cells which are found in the majority of the species. In *Bazzania*, on the other hand, the power of developing terminal branches from ventral segments was lost, and the flagelliform branches which had arisen in this way in the ancestral stock were entirely replaced by intercalary flagelliform branches. At the same time the generalized type of structure in the axial orgaus and the bifid character of the leaves were retained in the more primitive members of the genus. In the more typical species, however, the bifid leaves gave way to tridentate leaves.

The genus *Lepidozia* may possibly have been derived from the same ancestral stock. If this was the case terminal branching from the ventral segments disappeared, although the power of giving off such branches was retained by the lateral segments. The ventral flagelliform branches, as in *Bazzania*, became intercalary in orgin, but their number was greatly reduced. They were replaced, to a certain extent, by lateral vegetative branches, the tips of which acquired a flagelliform character. The genus gives further evidence of advance in the differentiation of its axial tissues into cortex and medulla and in its plurifid leaves, which have almost entirely replaced the bifid leaves of the more primitive types. The pinnate appearance of the branch-systems, which is so striking in certain species, is brought about by the limited growth of the lateral branches and by their subordinate position with respect to the higher axes. This too is evidence of advance, because it indicates a higher degree of differentiation in the branches. In the apparently dichotomous branch-systems of *Acromastigum* and *Bazzania* the lateral branches are unlimited in growth and essentially like the higher axes from which they spring.

The genera *Acromastigum*, *Bazzania*, and *Lepidozia* form a well-defined group within the *Cephaloziaceae* (or *Trigonantheae*). With this group might be associated, as outlying members, the African genus *Sprucella* and the tropical American genus *Micropterygium*. All five genera are characterized by incubous leaves, relatively large under-leaves, flagelliform branches, and short, well-differentiated sexual branches, which are ventral in position and intercalary in origin. In *Sprucella*, which is much like *Lepidozia*, terminal branches from lateral segments are present, and the underleaves are definitely quadrifid; the leaves, however, are bidentate rather than plurifid. In *Micropterygium*, which is more remote, the leaves are complicate and develop a wing along the keel (see REIMERS, 26). Both lateral and ventral vegetative branches, moreover, are intercalary, and the power of producing terminal branches is apparently completely lost. The widely distributed genus *Calypogeia* might also be included in the group, although the marsupia formed by the female branches make its systematic position open to question. In *Calypogeia* the leaves are incubous and the underleaves relatively large, but lateral branches seem to be completely absent and all the ventral branches are intercalary in character.

The other genera of the *Cephaloziaceae*, especially those allied to *Cephalozia*, are very distinct from the group under consideration. Most of these genera are characterized by succubous leaves and by minute or obsolete underleaves, although a few develop underleaves of a relatively large size. It is possible that further study may justify the division of the *Cephaloziaceae* into two or more distinct families.

GEOGRAPHICAL DISTRIBUTION

Our knowledge regarding the geographical distribution of the various species of *Acromastigum* is far from satisfactory. Of the twenty-eight species recognized in the present revision the known range, in the case of twenty-one, is restricted to a single country or island. Ten of these species, in fact, are known only from the type-collections. Since many of the species are small, inconspicuous, and easily overlooked, it is probable that undescribed species still await discovery, and that the species already described will be found in new localities, perhaps far removed from their present known ranges.

According to the information available the island of Borneo is the region in which the genus is best represented. Ten species have been found on the island and six of these are known from nowhere else. The peninsula of Malacca, with five species to its credit, comes next in order of abundance. This is followed by Amboina, Australia, Banka, Chile, New Caledonia, New Zealand, the Philippines, and Sumatra, with three species apiece, and by Java, with two species. The other regions where the genus occurs lie outside the main area of distribution and include Nepal, South Africa, and various islands of the Pacific. Each of these countries or islands has but a single species.

If Borneo represents the focal center for the genus the area of distribution spread out in various directions. One path, for example, apparently passed through Java, Sumatra and the peninsula of Malacca to Nepal; another extended in a northeasterly direction to the Philippines; a third reached as far as Chile, by way of Australia and New Zealand; and a fourth sent out branches through the Pacific, one of which terminated in Hawaii and another in Samoa. The occurrence of the genus in South Africa is an example of discontinuous distribution and is difficult to explain at the present time.

Seven species of *Acromastigum* have a known range extending beyond the limits of a single country or island. The most widely distributed of all is *A. anisostomum*, which is known from Australia, Chile, New Zealand, and Tasmania, as well as from several islands in the vicinity of New Zealand. The Indo-Malayan *A. inaequilaterum* also is widely distributed. The range of this species extends from Nepal and Malacca, through several East Indian islands, to Amboina and New Guinea. The other species of this group include four Indo-Malayan and one Australasian species. Of the Indo-Malayan species *A. divaricatum* is known from the Philippines, Malacca, and three East Indian islands; *A. echinatiforme* from the Philippines and three East Indian islands; *A. bancanum* and *A. echinatum* from Malacca, Banka, and Borneo. The Australasian species, *A. Colensoanum*, is known from Australia, New Zealand, and Tasmania.

LITERATURE CITED

The titles listed below are referred to in the body of the text; a few additional citations are included in the synonymy of the species.

1. ANDREAS, J. Ueber den Bau der Wand und die Oeffnungsweise des Lebermoos-sporogons. Flora 86: 161—213 f. 1—29, pl. 12. 1899.

2. BASTOW, R. A. Tasmanian Hepaticae. Papers & Proc. Roy. Soc. Tasmania 1887: 209—289. pl. 1—34. 1888.

3. BUCH, H. Die Scapanien Nordeuropas und Siberiens. I. Comm. Biol. Soc. Sci. Fennica 1⁴: 1—21. f. 1—11. 1933.

4. CARL, H. Über die vegetative Vermehrung in der Lebermoosgattung *Plagiochila*. Hedwigia 148—155. f. 1—4. 1932.

5. CARRINGTON, B., & PEARSON, W. H. List of Hepaticae collected by Mr. Thomas Whitelegge in New South Wales, 1884—5. Proc. Linn. Soc. New South Wales II. 2: 1035—1060. pl. 22—37. 1884.

6. COCKAYNE, L. The vegetation of New Zealand. In Engler & Drude, Die Vegetation der Erde 14: i—xxii, 1—364, pl. 1—65. Leipzig, 1921.

7. CORRENS, C. Untersuchungen über die Vermehrung der Laubmoose durch Brutorgane und Stecklinge. i—xxiv, 1—472. f. 1—87. Jena, 1899.

8. DE NOTARIS, G. Epatiche di Borneo raccolte dal D.ʳᵉ O. Beccari nel Ragiato di Sarawak durante gli anni 1865—66—67. Mem. R. Accad. Torino II. 28: 267—308. pl. 1—35. 1874.

9. DOUIN, CH. Le pédicelle de la capsule des Hépatiques. Bull. Soc. Bot. France 55: 194—202, 270—276, 360—366, 368—376. pl. 6—9. 1908.

10. EVANS, A. W. A new genus of Hepaticae from the Hawaiian Islands. Bull. Torrey Club 27: 97—104. f. A, B, pl. 1. 1900.

11. —— Branching in the leafy Hepaticae. Ann. Bot. 26: 1—37. f. 1—36. 1912.

12. —— Some representative species of *Bazzania* from Sumatra. Papers Michigan Acad. Sci., Arts and Letters 17: 69—118. pl. 13—18. 1933.

13. GOEBEL, K. Organographie der Pflanzen, insbesondere der Archegoniaten und Samenpflanzen. Erster Teil. Allgemeine Organographie. Dritte Auflage. i—ix, 1—642. f. 1—621. Jena, 1928.

14. GOTTSCHE, C. M., LINDENBERG, J. B. G., & NEES AB ESENBECK, C. G. Synopsis Hepaticarum, i—xxvi, 1—834. Hamburg, 1844—47.

15. HERZOG, TH. Anatomie der Lebermoose. In Linsbauer, Handb. Pflanzenanat. 7¹: 1—112. f. 1—93. 1925.

16. —— Geographie der Moose. i—xi, 1—439. pl. 1—8. Jena, 1926.

17. —— Hepaticae Philippinenses a cl. C. J. Baker lectae. Ann. Bryol. 4: 79—94, f. 1—4. 1931.

18. —— Beiträge zur Flora von Borneo. 11. Hepaticae. Mitteil. Inst. Allg. Bot. Hamburg 7: 182—216. f. 1—10. 1931.

19. LEITGEB, H. Untersuchungen ueber die Lebermoose. 2. Die foliosen Jungermannien. 1—95. pl. 1—12. Jena, 1875.

20. LINDENBERG, J. B. G., & GOTTSCHE, C. M. Species Hepaticarum. *Mastigobryum.* i—xii, 1—118, p. 1—22. Bonn, 1851.

21. MASSALONGO, C. Epatiche raccolte alla Terra del Fuoco dal Dott. C. Spegazzini nell' anno 1882. Nuovo Gior. Bot. Ital. 17: 201—217. pl. 12—27. 1885.

22. MITTEN, W. Hepaticae Indiae Orientalis: an enumeration of the Hepaticae of the East Indies. Jour. Linn. Soc. Bot. 5: 89—128. 1861.

23. —— Hepaticae. In Hooker, Handbook of the New Zealand Flora 497—549. London, 1867.

24. —— Jungermanniae. In Seemann, Flora Vitiensis 404—418. London, 1871.

25. MÜLLER, K. Die Lebermoose. Rabenhorst's Kryptogamen-Flora von Deutschland, Oesterreich und der Schweiz. Zweite Auflage. 6²: 1—947. f. 1—207. 1912—1916.

26. REIMERS, H. Revision der Lebermoosgattung *Micropterygium.* Hedwigia 73: 133—204. f. 1—20. 1933.

27. RODWAY, L. Tasmanian Bryophyta. Hepatics. Papers & Proc. Roy. Soc. Tasmania 1916: 1—92. 1916.

28. SANDE LACOSTE, C. M. VAN DER. Hepaticae. Jungermanniae Archipelagi Indici, adiectis quibusdam speciebus japonicis. Ann. Mus. Bot. Lugdano-Batavi 1: 287—314. pl. 7, 8. 1864.

29. SCHIFFNER, V. Lebermoose (Hepaticae) mit Zugrundelegung der von Dr. A. C. M. Gottsche ausgeführten Vorarbeiten. Forschungsreise „Gazelle" 4⁴: 1—48. pl. 1—8. 1980.

30. —— Ueber exotische Hepaticae, hauptsächlich aus Java, Amboina und Brasilien. Nova Acta Acad. Leop.-Carol. 60: 219—316. pl. 1—19. 1893.

31. —— Hepaticae (Lebermoose). In Engler-Prantl, Die natürlichen Pflanzenfamilien 1³: 3—141. f. 1—141. 1893.

32. —— Conspectus Hepaticarum Archipelagi Indici. 1—382. Batavia, 1898.

33. —— Untersuchungen über Amphigastrial-Antheridien und über den Bau der Andröcien der Ptilidioideen. Hedwigia 50: 146—162.f. 1—39. 1910.

34. SPRUCE, R. Hepaticae Amazonicae et Andinae. Proc. Bot. Soc. [Edinburgh] 15: i—xi, 1—588. pl. 1—22. 1885.

35. STEPHANI, F. Hepaticarum species novae vel minus cognitae. VI. Hedwigia 25: 133—134. pl. 3—6. 1886.

36. —— Hepaticarum species novae vel minus cognitae. VIII. Hedwigia 25: 233—249. pl. 1, 2. 1886.

37. —— Colenso's New-Zealand Hepaticae. Jour. Linn. Soc. Bot. 29: 263—280. pl. 26—28. 1892.

38. —— Beiträge zur Lebermoos-Flora Westpatagoniens und des südlichen Chile. Bihang K. Svenska Vet.-Akad. Handl. 26³ (No. 6): 1—60. 1900.

39. —— *Mastigobryum* Nees 1844. Spec. Hepat. 3: 413—540. 1908, 09. Pages 413—516 were reprinted from Bull. Herb. Boissier II. 8: 681—696, 745—776, 837—866, 941—966.

40. —— Botanische Ergebnisse der schwedischen Expedition nach Patagonien und dem Feuerlande 1907—1909. II. Die Lebermoose. Kungl. Svenska Vetensk.-Akad. Handl. 46: 1—92. f. 1—35. 1911.

41. —— *Mastigobryum*, Nees. Spec. Hepat. 6: 452—489. 1924.

42. VERDOORN, FR. De Levermosgeslachten van Java en Sumatra. Nederl. Kriudk. Archief 1931: 461—509. f. 1—93. 1931.

INDEX

Some Reviews of the "Manual of Bryology":

The editor of the "Annales Bryologici" has produced in this volume what is practically a general text-book of Bryology. Bryology like other branches of botany, suffers from lack of co-ordination among its workers: those occupied in general research sometimes have very little knowledge of the plants which with they deal and the value of their work is lessened by this narrowness of outlook. The taxonomists on the other hand, do not pay enough attention to general botanical research on the group; hence the ill-founded opinions and erroneous conclusions in many otherwise sound taxonomic papers. The present manual is an attempt to meet some of the difficulties. The book which is clearly printed and admirably produced forms a valuable addition to general botanical literature. *„Journal of Botany."*

Ein Handbuch der Bryologie und in dieser Art etwas völlig Neues!
.... darf sich rühmen, die bryologische Literatur um ein ebenso neuartiges wie zuverlässiges, und zudem in textlich wie illustrativer Hinsicht hervorragend ausgestattetes Werk bereichert zu haben.
„Die Naturwissenschaften".

Dette Vaerk kan betegnes som en „Almindelig Botanik" eller en „Biologi" vedrörende Mosserne alene. Den omhandler i 16 Afhandlinger, udarbejdede af 14 Forfattere, Mossernes almindelige Forhold. Alt nyt og alle nye Synspunkter er kommet med, og Bogen fremtraeder i alle Henseender som en fuldt ud moderne Haandbog. *„Botanisk Tisdkrift".*

Ongetwijfeld voorziet dit handboek in een dringende behoefte voor allen, die zich in de bryologie willen oriënteeren. Zooals de titel reeds aanduidt, maakt het aanspraak op de naam handboek en er is dan ook vrijwel geen onderdeel onbesproken gebleven. De redacteur heeft zich op bewonderenswaardige wijze van de hulp van talrijke vooraanstaande bryologen kunnen verzekeren. *„Vakblad voor Biologen".*

Con la pubblicazione di questo manuale il Verdoorn ha provveduto a fornire la letteratura botanica di un'opera importante, di cui si sentiva grande bisogno. Raccogliere quello che era già noto sui Muschi e sulle Epatiche e aggiornarlo con le recenti, interessanti ricerche fatte nei vari rami della Briologia, era lavoro tutt' altro che semplice; il Verdoorn lo ha potuto compiere felicemente assicurandosi la collaborazione di valenti specialisti.
„Annali di Botanica".

The Annales Bryologici are not devoted to taxonomy alone, but each volume also contains original contributions on general bryology. Monographs and memoirs, of a nature to form a complete work, are published in supplementary volumes. See next page.